The Age of Adaptation

How Climate Change is Reshaping Our World and Our Minds

David Collins

The Age of Adaptation

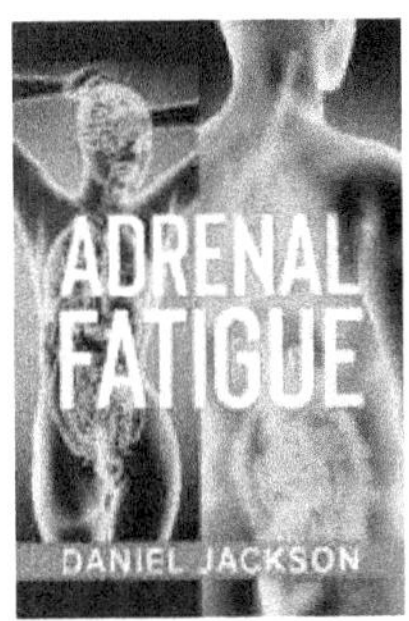

Take a look at more great books available from
Rockwood Publishing

... some for FREE!

Just visit the link below:

rockwoodpublishing.co.uk

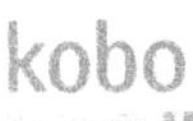

Contents

Chapter 1: The Unavoidable Truth: Climate Change is Here

The Manifestations of Climate Change: From Rising Temperatures to Rising Seas

Take a moment to think about your favorite place in nature. Maybe it's a sandy beach that hugs the edge of the ocean, a serene forest where sunlight filters through the leafy canopy, or perhaps a majestic mountain range that reaches up to touch the sky. No matter the location, all these spots share one ominous reality: they are being reshaped, right now, by the inexorable forces of climate change.

Climate change, once a distant threat relegated to the realm of abstract scientific models and debates, is now an unavoidable truth. It's like a specter that has been creeping up on us, unheeded by many, yet never pausing in its relentless advance. Suddenly, it's no longer creeping—it's running, and it has overtaken us. We are in the age of climate change. Not on the cusp, not on the brink—we are living it.

Understanding this reality, let's delve into the multitude of ways climate change manifests. The two most prominent signs are temperature increases and sea level rise, both of which have far-reaching consequences.

Rising Temperatures

Climate change is most often associated with rising global temperatures—hence the term 'global warming'. The Earth's average surface temperature has risen by about 1.2 degrees Celsius since the late 19th century. While this number might seem small on the surface, remember, it's an average increase across the entire planet. This means there are regions where the temperature increase is far more drastic. In the Arctic, for instance, warming has been twice as fast as the global average, a phenomenon known as 'Arctic amplification'.

This overall warming trend affects more than just our sense of comfort on a hot summer day. Ecosystems, both on land and in the ocean, are delicately balanced. Even a small increase in temperature can upset this balance. We're already witnessing this in the form of devastating forest fires, more frequent and extreme heatwaves, and species migrations to cooler regions. And this is just the tip of the melting iceberg.

Rising Seas

Sea-level rise is another undeniable manifestation of climate change. As the planet warms, two things happen: glaciers and ice sheets melt, adding more water to the ocean, and the ocean itself expands as it warms, a process known as thermal expansion. Both of these processes contribute to a rising sea level.

So far, sea levels have risen by about 8 inches since reliable record-keeping began in the late 19th century. Projections for the future, however, are more alarming. The latest reports suggest that we could see a rise of up to 1.3 meters by the end of this century if greenhouse gas emissions continue unchecked.

This rise in sea level has profound implications for coastal communities. It leads to more frequent and severe coastal flooding, increased coastal erosion, and—in some low-lying regions—it threatens to submerge land entirely. Iconic cities like Venice and Miami are already grappling with these realities, investing billions in mitigation strategies.

The manifestations of climate change are not limited to just rising temperatures and rising seas. We are also seeing more frequent and intense extreme weather events, changes in precipitation patterns leading to alternating periods of drought and intense rainfall, ocean acidification, and biodiversity loss. And these changes interact with each other in complex ways, often exacerbating the impacts of each individual factor.

So, the age of adaptation is upon us. For decades, we've been warned about the threat of climate change, but now it's here and it's tangible. We need to face this unavoidable truth and understand its many manifestations. But more than just understanding, we need to act.

Extreme Weather Events

Another significant manifestation of climate change is the increase in the frequency and intensity of extreme weather events. Tornadoes, hurricanes, heatwaves, and heavy rainfall events have all become more frequent and severe in recent years. You might remember 'once-in-a-century' storms – well, they're no longer a rarity, but rather a recurring nightmare in many parts of the world.

For example, hurricanes, powered by warm ocean water, have become more potent. A warmer atmosphere can hold more moisture, leading to heavier rainfall during these storms, while sea-level rise increases the risk of storm surges and coastal flooding.

Changing Precipitation Patterns

Climate change doesn't just bring about more rain; it also changes when and where this rain falls. Some regions are experiencing longer periods of drought, interspersed with episodes of intense rainfall. These altered patterns can lead to both water shortages and flooding, posing serious challenges to agriculture, water management, and disaster preparedness.

Ocean Acidification

Our oceans have been acting as a buffer, absorbing about 30% of the CO2 we emit, thus mitigating the rate of global warming. But this comes at a severe cost. When CO2 dissolves in seawater, it forms carbonic acid, leading to ocean acidification. This change in the ocean's chemistry has a drastic impact on marine life, especially creatures that form shells or skeletons, like corals and shellfish. These organisms struggle to form their calcium carbonate structures in more acidic conditions, threatening the survival of these species and the larger marine ecosystems they support.

Biodiversity Loss

From the tiniest insects to the largest mammals, climate change affects all life on Earth. Altered temperatures and precipitation patterns, combined with extreme weather events, are transforming habitats and forcing species to adapt, move, or perish. Polar species, like the polar bear, are perhaps the most iconic victims of this phenomenon, with their icy habitats literally melting beneath their feet.

But even species in more temperate regions are affected. Changes in flowering times or shifts in available prey can disrupt finely tuned ecological relationships, leading to declines or even extinctions. This loss of biodiversity could have profound implications, potentially unraveling the complex web of life that supports us all.

Climate change is not a far-off threat; it is a present and all-encompassing reality. It's not just about warmer summers and milder winters. It's about fundamental shifts in the systems that sustain life on Earth. It's about the rise of seas and the fall of species. It's about more intense storms and more frequent droughts.

But let's not forget, it's also about us—humans. It's about our cities threatened by rising seas, our farms parched by drought, our homes battered by storms. But more importantly, it's about how we respond. Do we despair, or do we adapt? The choice is ours, and the time to make that choice is now. In the age of adaptation, our actions (or inactions) will determine not only our future but the future of the countless species we share this planet with.

In the chapters that follow, we'll explore these themes more deeply, providing a clear and comprehensive picture of our changing world. From the thawing Arctic to the bleaching coral reefs, from the burning forests to the flooded coastlines—we'll journey through our transforming world, understanding its changes and exploring our potential to adapt. So, buckle up; it's going to be a challenging, enlightening, and hopefully inspiring journey.

Major Recent Climate Events and Their Impact

While statistics and global trends serve to illustrate the overarching narrative of climate change, there is something distinctly poignant about individual events. They bring the broader picture into sharp, often tragic, focus. Let's turn our attention now to some of the major climate events of recent years, which demonstrate the urgency of the crisis at hand.

Heatwaves and Forest Fires

In recent years, we've seen heatwaves of unprecedented intensity and duration. In the summer of 2020, the U.S. experienced one of its most extensive heatwaves, with more than 50 million people under heat advisories. The same year, Siberia reported a record-breaking 38 degrees Celsius (100.4 degrees Fahrenheit), a figure that would be unusual even in many tropical regions, let alone within the Arctic Circle.

Heatwaves are more than an uncomfortable inconvenience; they have severe implications for human health and infrastructure. High temperatures exacerbate respiratory and cardiovascular problems, lead to heatstroke, and can be fatal, especially for the young, the elderly, and those with underlying health issues. Additionally, heatwaves strain our energy grids due to increased use of air conditioning, sometimes leading to blackouts.

Moreover, intense heat provides fertile ground for forest fires. The wildfires that swept through Australia at the end of 2019 and beginning of 2020, known as the Black Summer bushfires, serve as a harrowing example. They burned an estimated 18.6 million hectares, destroyed over 5,900 buildings (including 2,779 homes), and killed at least 34 people. Beyond the human tragedy, the ecological loss was staggering: An estimated 1 billion animals perished, and several species were pushed to the brink of extinction.

Similarly, California has been besieged by increasingly severe fire seasons. The 2020 fire season was the largest recorded in California's history, with over 4 million acres burned, dwarfing previous records. Towns were razed, lives were lost, and the state was blanketed with smoke, leading to some of the worst air quality in the world.

Floods and Cyclones

On the other end of the spectrum are floods. In 2021, China experienced one of its worst flooding events in recent history. Persistent heavy rains led to severe flooding in Henan province, affecting over 1.4 million people, destroying homes, and damaging infrastructure. The city of Zhengzhou was particularly hard hit, with a year's worth of rain falling in just three days.

Such extreme rainfall events are becoming more common. As global temperatures rise, the air can hold more moisture, leading to heavier rainfall during storms. When this increased precipitation combines with rising sea levels, the result is more frequent and intense flooding, especially in coastal areas.

This was also tragically exemplified by Cyclone Amphan, which struck Bangladesh and India in May 2020. As one of the most powerful cyclones ever recorded in the Bay of Bengal, it brought widespread destruction, storm surges, and heavy rainfall. Over 90 people lost their lives, and millions were displaced. And yet, the death toll could have been much worse. Lessons from previous cyclones had led to better-prepared disaster response systems, saving countless lives.

Melting Ice and Rising Seas

Perhaps no place better embodies the impacts of rising temperatures than the Arctic. The region is warming twice as fast as the rest of the planet, a phenomenon known as Arctic amplification. The consequences of this rapid warming are starkly evident in the disappearing sea ice.
In 2020, Arctic sea ice shrank to its second-lowest extent on record. This not only threatens polar species but has global implications. Sea ice reflects sunlight back into space. When it melts, it exposes the dark water underneath, which absorbs more heat, further amplifying global warming—a phenomenon known as the albedo effect.

Meanwhile, the world's ice sheets are also in peril. Greenland, home to the largest ice sheet outside Antarctica, is losing ice at an alarming rate. In 2019, it shed an unprecedented 600 billion tons, contributing to a global sea-level rise of 1.5 millimeters that year alone.

Rising seas have direct implications for low-lying coastal areas and islands. The Pacific Island nation of Tuvalu, with its highest point only 4.5 meters above sea level, is acutely at risk. Saltwater intrusion into its freshwater supply, increased coastal erosion, and the ever-present threat of being submerged are realities the Tuvaluans live with every day.

Climate Change and Public Health

Finally, it's essential to mention the ongoing COVID-19 pandemic. While it's not a climate event, the parallels and intersections with climate change are many and significant. Both are global crises, disproportionately affecting the most vulnerable and requiring a coordinated international response. Furthermore, the prevention of future pandemics depends on addressing climate change. Warmer temperatures and habitat loss can alter the spread of vector-borne diseases and bring humans into closer contact with wildlife, increasing the risk of zoonotic diseases.

The impacts of climate change are not just future predictions—they're unfolding here and now. The heatwaves scorching our lands, the fires consuming our forests, the floods submerging our towns, the cyclones battering our coasts, and the ice sheets melting into our oceans, are all somber reminders of this reality.

Unprecedented Heatwaves

The year 2021 witnessed extraordinary heatwaves. Canada's westernmost province, British Columbia, suffered a severe heatwave in late June and early July, causing the deaths of hundreds of people and triggering wildfires. The village of Lytton broke Canada's highest-ever recorded temperature, reaching a sweltering 49.6°C (121.3°F) on June 29.

This record-breaking heat event wasn't isolated. It was part of a more extensive heat dome event that covered much of western North America, including the U.S. Pacific Northwest. Cities known for their cool and mild climate, such as Portland and Seattle, recorded temperatures above 40°C (104°F), shattering previous records.

Increased Cyclone Activity

In the Southern Hemisphere, the 2021-2022 cyclone season was particularly intense, especially in the South Pacific and South Indian Ocean. Tropical cyclones Ana and Batsirai wreaked havoc in Madagascar and Mozambique, leading to significant loss of life and infrastructure damage. These events underscored the vulnerability of many low-lying coastal communities to the increased intensity of tropical cyclones in a warming world.

Extreme Rainfall and Flooding

Across the globe, extreme rainfall events led to widespread flooding. In July 2021, Western Europe experienced a devastating flood event when regions in Germany, Belgium, and the Netherlands received up to two months' worth of rainfall in just two days. Entire towns were destroyed, with over 200 lives lost in what was one of the worst natural disasters in the region in recent memory.

In August 2021, the remnants of Hurricane Ida dumped a record-breaking amount of rain on the U.S. Northeast, causing flash floods that resulted in dozens of fatalities. New York City saw its highest-ever recorded rainfall rate, with nearly 80 mm falling in just one hour.

Arctic Ice Melt

In the far north, Arctic sea ice continued its troubling decline. The summer of 2022 saw the Arctic sea ice extent fall to its second-lowest level on record, reflecting the region's ongoing and rapid warming. This loss of sea ice threatens Arctic wildlife, including polar bears and seals, and contributes to global sea-level rise.

Drought and Water Scarcity

Conversely, many regions worldwide faced severe droughts. The American West, from California to Colorado, grappled with one of its worst droughts in centuries in 2022. Reservoirs dropped to critically low levels, leading to water shortages and agricultural losses, and heightening the risk of wildfires.

Biodiversity Loss

The year 2023 was marked by a concerning report from the Intergovernmental Science-Policy Platform on Biodiversity and Ecosystem Services (IPBES). It revealed an acceleration in biodiversity loss, with many species edging closer to extinction due to climate change, habitat loss, and other human-induced pressures.

Each of these events underscores the reality that climate change is not a distant problem, but a present crisis. The good news is that awareness of the climate crisis has never been higher. Countries, cities, businesses, and individuals are taking action, from transitioning to renewable energy to implementing more sustainable practices. As we explore in the next chapters, this collective action gives us hope for a more sustainable, resilient future.

However, amidst this bleak narrative, there are also stories of resilience and adaptation. From improved disaster response saving lives in the face of cyclones to efforts by the Tuvaluans to 'live with the water,' we are learning to navigate this new climate reality. As we move forward, these lessons in resilience and adaptation will become increasingly important.

Decoding the Science: The Causes and Predicted Trajectories of Climate Change

To navigate the future and understand our role within it, we need to understand the underpinnings of climate change. This journey requires a look at the causes of climate change, how we model its future, and what these predictions mean for our world. Let's embark on this exploration to decode the science of climate change.

Causes of Climate Change

Climate change is a naturally occurring phenomenon, with Earth's climate evolving over millions of years due to factors like volcanic activity, changes in the Earth's orbit, and variations in the sun's energy. However, what we're currently experiencing is anything but natural. This unprecedented rate of change, largely occurring over the last century, is predominantly driven by human activities – hence the term 'anthropogenic climate change.'

The main culprit is the increase in greenhouse gases (GHGs) in the Earth's atmosphere. GHGs, which include carbon dioxide (CO_2), methane (CH_4), and nitrous oxide (N_2O), trap heat from the sun, creating a 'greenhouse effect' that warms the planet. This process is essential for life as we know it; without it, our world would be too cold to sustain life. However, the problem arises when the concentration of these gases increases.

Since the Industrial Revolution, human activities, particularly the burning of fossil fuels (coal, oil, and gas) and deforestation, have significantly increased the concentration of GHGs in the atmosphere. For instance, the amount of CO_2 in the atmosphere now far exceeds the natural range seen over the last 800,000 years. This surplus of GHGs enhances the greenhouse effect, leading to a rapid and unnatural warming of the planet - known as global warming.

Climate Models and Predictions

Understanding the future trajectory of climate change involves complex models that consider a myriad of variables, from GHG emissions to cloud cover to ocean currents. Climate models are computer programs that simulate the interactions of the atmospheric, oceanic, terrestrial, and cryospheric systems. They're our best tool for predicting future climate change.

However, the future isn't set in stone; it depends on our actions. Therefore, climate models typically consider different emission scenarios, ranging from 'business-as-usual' scenarios, where GHG emissions continue to rise at the current rate, to more optimistic scenarios where significant mitigation measures are implemented.

These models have been remarkably accurate, with predictions made decades ago aligning closely with the changes we're witnessing today. So, what do these models tell us about our future?

Predicted Trajectories

If GHG emissions continue unabated, global average temperatures could increase by 4-5°C by the end of the century, according to the Intergovernmental Panel on Climate Change (IPCC). This level of warming would have catastrophic impacts, including severe heatwaves, increased frequency and intensity of extreme weather events, significant sea-level rise, and drastic ecological disruptions.

However, if we manage to limit global warming to 1.5°C, a target set by the Paris Agreement, these impacts could be substantially reduced. Achieving this goal would require rapid and far-reaching transitions in energy, land use, infrastructure, and industrial systems.

One aspect of these models that can be alarming is the concept of 'tipping points.' These are thresholds in the climate system that, once crossed, can lead to irreversible changes. Examples include the loss of the Amazon rainforest or the melting of the Greenland ice sheet. The exact thresholds for these tipping points are uncertain, making them a subject of intense research and a significant concern.

A Changing World

Regardless of the scenario, some degree of warming is locked in due to the GHGs already in the atmosphere. As a result, we will have to adapt to a changing world. Even in the best-case scenario, we will likely witness the disappearance of some low-lying islands due to sea-level rise, shifts in agricultural patterns due to changing rainfall and temperatures, and an increase in heatwaves and extreme weather events.

Adapting to these changes will require resilience and innovation. Some changes can be straightforward, such as developing heat-resistant crops or improving infrastructure to withstand extreme weather. Others, like dealing with significant sea-level rise or shifting ecosystems, will be more challenging and may require more transformative changes.

The science also underscores the importance of 'climate justice.' The impacts of climate change are not evenly distributed. Those who have contributed the least to the problem—often developing countries and marginalized communities—are the most vulnerable to its impacts. This raises important questions about equity and justice in responding to climate change, a theme we will explore further in subsequent chapters.

The Future is in Our Hands

While these predictions paint a sobering picture, it's essential to remember that the future is not yet written. Our actions today will shape the world of tomorrow. Decarbonizing our economies, transitioning to renewable energy, protecting and restoring forests, and adapting to changes are all within our reach.
One encouraging development is the falling cost of renewable energy. Solar and wind power are now among the cheapest sources of electricity in many parts of the world, making the transition to clean energy not only environmentally necessary but also economically attractive.

Similarly, we're seeing growing recognition of the importance of nature in tackling climate change. Forests, wetlands, and other ecosystems absorb vast amounts of CO_2 and are crucial for preserving biodiversity. Protecting and restoring these 'natural climate solutions' can provide a third of the climate mitigation needed by 2030 to keep warming below 2°C.

Finally, there's growing awareness and action on climate change at all levels, from individuals to governments. More and more countries are setting net-zero emission targets, businesses are investing in sustainability, and young people are mobilizing for climate action.

While the science of climate change warns us of the risks and challenges ahead, it also illuminates the path forward. It empowers us with the knowledge we need to shape a sustainable, resilient future.

In the next chapters, we'll delve into this journey, exploring the solutions and opportunities that lie ahead in the age of adaptation.

Chapter 2: The Psychological Landscape of Climate Change

The Concept of Eco-Anxiety: Understanding Our Fear

As we delve further into the age of adaptation, it is important to consider not just the physical impacts of climate change, but also the psychological. Our minds are not separate from our environment, and the profound alterations occurring around us are reshaping our mental and emotional landscapes. One notable manifestation of this mental-emotional shift is eco-anxiety.

Eco-anxiety, or climate anxiety, is a relatively new term that describes the feelings of fear, distress, and concern triggered by the environmental crisis. It's an existential dread, knowing that our future, and more significantly, the future of generations to come, is threatened. The American Psychological Association defines eco-anxiety as "a chronic fear of environmental doom."

Eco-anxiety arises from the understanding of what we stand to lose—our comfortable lifestyles, our familiar landscapes, perhaps even our lives—and the uncertainty of what's to come. It is compounded by the feeling of helplessness, of being trapped in systems that seem beyond our control. Whether it's melting ice caps or unprecedented wildfires, the constant stream of apocalyptic news can be overwhelming, amplifying these fears.

This anxiety is not merely an abstract concept. It has real, tangible effects on people's well-being. Those experiencing eco-anxiety can exhibit symptoms similar to those of other anxiety disorders, including restlessness, fatigue, concentration problems, irritability, sleep disruption, and feelings of despair.

Eco-anxiety is not evenly distributed. Just as the physical impacts of climate change disproportionately affect certain communities, so too does eco-anxiety. Young people, who will bear the brunt of climate change, are particularly susceptible. A 2019 survey by the charity Global Action Plan found that 70% of 14 to 24-year-olds in the UK believe their mental health is being affected by the threat of climate change.

Those on the front lines of the climate crisis are also significantly affected. Farmers facing droughts, inhabitants of low-lying islands at risk of submersion, and residents of wildfire-prone areas live with the constant fear and stress of losing their homes, livelihoods, and communities. This anxiety can contribute to a host of mental health problems, including depression and post-traumatic stress disorder (PTSD).

Eco-anxiety can also influence our behavior and decision-making. It can deter people from having children, for fear of the world they'll inherit. It can prompt lifestyle changes, from reducing carbon footprints to engaging in climate activism. And in some cases, it can lead to 'climate paralysis,' a state of feeling so overwhelmed that one becomes apathetic and inactive.

Climate Change and Cognitive Dissonance: Ignorance is Bliss?

While eco-anxiety represents an intense, conscious response to the climate crisis, other psychological phenomena can lead us in the opposite direction: towards denial, inaction, or apathy. One such phenomenon, integral to our understanding of the human response to climate change, is cognitive dissonance.

What is Cognitive Dissonance?

Cognitive dissonance, a theory introduced by psychologist Leon Festinger in the 1950s, refers to the mental discomfort experienced when a person holds two or more contradictory beliefs, values, or attitudes simultaneously, or is confronted with new information that contradicts existing beliefs, values, or attitudes. It is a state of tension that motivates us to change our beliefs or behaviors to reduce this discomfort.

For example, someone who cares deeply about the environment (belief) but flies frequently for work (behavior) may experience cognitive dissonance. To alleviate this tension, they may change their behavior (e.g., fly less) or modify their belief (e.g., convince themselves that their personal contribution to emissions is negligible).

Cognitive Dissonance and Climate Change

In the context of climate change, cognitive dissonance is commonplace. We live in societies largely built on fossil fuels, and our lifestyles are often at odds with the need to reduce emissions. We understand that driving cars contributes to CO_2 emissions, yet we continue to use them. We know excessive consumption and waste contribute to environmental degradation, yet we continue to buy and discard at unprecedented rates.

The cognitive dissonance resulting from this contradiction between understanding the reality of climate change and continuing to live in ways that contribute to it can manifest in several ways:

1. **Denial:** The most straightforward way to resolve cognitive dissonance is to reject the uncomfortable information. Climate change denial is a prime example of this. By denying the reality or severity of climate change, individuals can continue their behaviors without discomfort.
2. **Minimization:** Some accept the reality of climate change but downplay its severity or their role in it. This stance allows them to acknowledge the science without feeling obliged to change their behavior.
3. **Apathy:** Another response is to accept the reality and severity of climate change but to feel so

overwhelmed by the scale of the problem that they succumb to a sense of helplessness and disengage.

4. **Projection:** People may project the responsibility onto others - other individuals, corporations, or governments - thereby absolving themselves of the need to act.

5. **Action:** Finally, some choose to reduce dissonance through action, whether it's changing personal behaviors, advocating for policy changes, or engaging in climate activism.

The Role of Societal Structures

While cognitive dissonance is a personal experience, it's important to remember that it exists within broader societal structures. Our economic system, infrastructure, and cultural norms often make sustainable choices challenging or inconvenient. For instance, in many places, it's difficult to choose public transit over a personal car due to inadequate infrastructure. Similarly, the pressures of consumerism encourage excessive consumption and waste.

The Bliss of Ignorance?

The notion of 'ignorance is bliss' comes into play when the discomfort of cognitive dissonance leads to denial, minimization, or apathy. But is this 'bliss' truly blissful? While these strategies might offer temporary relief, they hinder our collective ability to address climate change and adapt to its impacts. In the long run, they harm not only the planet but also our own well-being.

Ultimately, the key to navigating cognitive dissonance in the age of climate change is not ignorance but awareness - awareness of our emotions, our contradictions, and the societal structures that shape our behaviors. By facing our discomfort and using it as a catalyst for change, we can move towards a more sustainable and resilient future.

Moving Beyond Dissonance: The Path to Personal and Collective Action

Resolving cognitive dissonance in the context of climate change is not a simple task. It often requires profound changes in our beliefs, attitudes, and behaviors - changes that can feel uncomfortable, daunting, or even threatening. However, many strategies can help us navigate this process and turn our discomfort into action:

1. **Education:** Knowledge is power. Understanding the science of climate change, the impacts of our actions, and the ways we can mitigate and adapt to climate change can empower us to make informed choices.

2. **Reflection:** Self-reflection is a critical tool for exploring our contradictions and identifying ways to align our actions with our values. Reflecting on questions such as "How do my actions contribute to climate change?" or "What changes am I willing and able to make?" can guide us towards more sustainable behaviors.

3. **Mindfulness:** Mindfulness practices, such as meditation, can help us sit with the discomfort of cognitive dissonance without rushing to easy resolutions. They can foster a sense of calm and clarity that allows us to confront the reality of climate change with resilience and compassion.

4. **Community:** Engaging with others - whether through climate activism, community initiatives, or simply conversations with friends and family - can alleviate feelings of overwhelm and helplessness. Collective action can also amplify our individual efforts and foster a sense of hope and solidarity.

5. **Advocacy:** Pushing for systemic changes is a vital part of resolving cognitive dissonance. By advocating for policies that make sustainable choices easier and more accessible, we can help bridge the gap between our values and our actions.

6. **Self-compassion:** It's important to remember that we're all navigating cognitive dissonance in some form, and it's okay to feel overwhelmed. Being gentle with ourselves, celebrating small victories, and remembering that every step counts can sustain us on this journey.

In this age of adaptation, cognitive dissonance can be both a challenge and an opportunity. It invites us to confront the contradictions of our modern lifestyles, to grapple with the uncomfortable truth of climate change, and to emerge with a renewed commitment to our planet and our future. As we navigate this psychological landscape, we'll discover that awareness, not ignorance, is our true bliss.

The Impact on Mental Health: Anticipated Developments

The challenges posed by climate change are not restricted to the environmental sphere alone; they transcend into the realm of our mental health, creating a burgeoning public health concern that the world must confront. In this section, we examine how these impacts are expected to develop and amplify, highlighting the urgency of integrating climate change and mental health initiatives.

Direct Impacts on Mental Health

The repercussions of climate change – severe weather events, displacement, resource scarcity – can lead to direct psychological trauma. These experiences are traumatic not only physically but also psychologically, often resulting in Post-Traumatic Stress Disorder (PTSD), anxiety, depression, and substance abuse disorders.

As climate change exacerbates, we can expect an escalation in such climate-related traumas, leading to increased incidence of mental health disorders. Areas prone to extreme weather events and with inadequate support infrastructure will be most affected, resulting in a landscape where mental health issues could become the norm rather than the exception.

Indirect Impacts on Mental Health

Besides the direct trauma induced by climate change events, there are subtle, indirect ways that the crisis affects our mental health. Changes in local climate can disrupt our connection to place, leading to loss of identity and an ensuing sense of helplessness and grief. This phenomenon, termed 'solastalgia,' is an emerging area of concern.

The chronic stress of living in a world increasingly defined by climate uncertainty can also take a toll on our mental wellbeing. Eco-anxiety, a term we've previously examined, refers to the persistent worries about the future in the context of climate change. As our climate continues to destabilize, such chronic stress responses are expected to become more prevalent, potentially leading to a rise in related conditions like anxiety disorders and depression.

Vulnerable Populations

Certain populations are more susceptible to the mental health effects of climate change. Individuals with pre-existing mental health conditions may see their symptoms exacerbated by climate-related stressors. Young people, who are growing up in an age of climate uncertainty, are particularly vulnerable to developing anxiety and depression. Climate refugees, or those displaced due to climate change, also face significant mental health risks due to the trauma of displacement and the challenges of adaptation.

Potential Development Pathways

As we look to the future, several potential development pathways emerge regarding the impact of climate change on mental health.

1. **Increase in mental health disorders:** If climate change continues unchecked, we can expect a rise in both acute and chronic mental health conditions. Direct trauma from extreme weather events will contribute to an uptick in acute disorders like PTSD, while the chronic stress of living with climate uncertainty will contribute to an increase in conditions like anxiety and depression.

2. **Greater inequities:** Mental health impacts, like the physical impacts of climate change, are likely to disproportionately affect marginalized populations. Those living in poverty, in regions with high climate risk, or with pre-existing mental health conditions will face greater mental health risks. This could deepen existing health inequities.

3. **Increased demand for mental health services:** The rise in mental health disorders will lead to increased demand for mental health services. Health systems, many of which are already under-resourced, may struggle to meet this increased demand, leading to a mental health care crisis.

4. **Rise in climate-related grief and loss:** As climate change progresses, more people will experience loss - loss of homes, loss of natural spaces, loss of a sense of security. This is likely to

result in an increase in climate-related grief and loss, necessitating new approaches to grief counseling and therapy.

While these developments paint a daunting picture, it's crucial to remember that this is not our predetermined fate. We have the tools and knowledge to mitigate these impacts. By taking action on climate change, improving our mental health infrastructure, and creating supportive communities, we can change this trajectory.

Resilience and Adaptation

While the impacts of climate change on mental health are significant, humans have an innate capacity for resilience and adaptation. Community resilience, in particular, can play a crucial role in mitigating mental health impacts. By fostering social cohesion and community support, we can buffer the mental health impacts of climate events and help individuals recover more quickly.

Furthermore, the concept of 'post-traumatic growth' suggests that adversity can lead to personal growth and psychological development. While this does not negate the suffering caused by climate events, it suggests that we can find ways to grow and learn from these experiences.

Policy Implications

The anticipated developments in climate change's mental health impacts have significant policy implications. They underscore the need for mental health to be a key component of climate adaptation strategies. This includes investing in mental health infrastructure, integrating mental health into disaster response plans, and implementing policies to mitigate climate change and its impacts.

In addition, we must prioritize mental health research to better understand these impacts, identify vulnerable populations, and develop effective interventions. This includes research on the mental health benefits of climate mitigation and adaptation actions, such as green spaces and community resilience initiatives.

The Power of Action

Finally, it's important to remember that action on climate change can itself be a powerful tool for mental health. Action can alleviate feelings of helplessness and anxiety, foster a sense of hope, and create a sense of purpose and meaning. Whether it's personal actions like reducing our carbon footprint, or collective actions like advocating for climate policy, each step we take not only contributes to the fight against climate change but also bolsters our mental well-being.

As we move further into the age of adaptation, it's clear that our mental landscape is changing alongside our physical one. The anticipated developments in the mental health impacts of climate change underscore the urgency of action. However, they also highlight our capacity for resilience and the power of action. By acknowledging and addressing these impacts, we can forge a path towards a mentally resilient future - a future where we not only survive but thrive, in a changing climate.

The Elephant in the Room: Is Climate Change a Natural Cyclical Occurrence?

Few topics inspire as much controversy as climate change, and one common argument posited by skeptics is that the warming we are witnessing is merely part of the Earth's natural climate cycle. As we delve into this subject, we aim to dissect this argument, examining the evidence for natural climate cycles and how they differ from the current scenario we find ourselves in.

The Earth's Climate: A Story of Change

The Earth's climate is indeed a story of change. Over the span of our planet's 4.5-billion-year history, it has seen periods of tropical warmth and icy cold. Ice ages and warm periods, known as interglacials, have punctuated the Earth's history, driven by a host of natural factors, including changes in the Earth's orbit, solar radiation, and volcanic activity.

For example, the Ice Ages, a term used to define periods of long-term reduction in the Earth's temperature, were primarily influenced by changes in the Earth's orbit and axis tilt, which affected the amount of sunlight reaching different parts of the Earth's surface. These orbital changes, known as Milankovitch cycles, occur on predictable timescales of tens of thousands to hundreds of thousands of years.

Likewise, volcanic activity can also influence the Earth's climate. Large-scale volcanic eruptions can eject significant volumes of gas and ash into the atmosphere, which can reflect sunlight and temporarily cool the Earth's climate.

So, given this evidence, is it not plausible that the current trends of global warming are just another chapter in this ongoing narrative of natural climate fluctuations?

Distinguishing Natural Cycles from Current Climate Change

While it is true that the Earth's climate has always been in flux, the current phase of rapid warming is an anomaly when viewed against the backdrop of natural climate variability. The key differences lie in the rate of change and the primary drivers.

1. **Rate of Change:** Perhaps the most striking difference between past natural climate changes and current global warming is the rate of change. Geological records show that the global temperature typically changed by 4-7 degrees Celsius over the course of tens to hundreds of thousands of years during the glacial-interglacial cycles. In stark contrast, human-induced global

warming has led to an increase of approximately 1 degree Celsius in just over a century, a rate of change at least ten times faster than any average rate of change in the last 65 million years.

2. **Primary Drivers:** As we've mentioned, natural climate changes in Earth's history were driven by factors such as orbital changes and volcanic activity. However, the current warming trend correlates closely with increasing concentrations of greenhouse gases in the atmosphere, primarily from human activities such as burning fossil fuels and deforestation. The Intergovernmental Panel on Climate Change (IPCC) states that it is "extremely likely" that more than half of the increase in Earth's average surface temperature since the mid-20th century is due to human influence.

In essence, while climate change can and does occur naturally, the present episode of global warming aligns closely with human activity, not natural cycles.

The Elephant Unmasked

The argument that climate change is just a 'natural cycle' is often used to downplay the need for action. However, this perspective neglects the unprecedented rate of current warming and the clear link to human activities. Our understanding of natural climate variations actually provides compelling evidence that human activity is disrupting our climate.

This is not to say that natural factors have ceased to influence the climate - they still play a role. But their influence pales in comparison to the potent impact of greenhouse gases that humanity is pumping into the atmosphere.

Addressing climate change requires that we face this elephant in the room and acknowledge our role in the unfolding crisis. By recognizing that the current warming is largely driven by human activities, we can begin to understand the need for immediate and substantial reductions in greenhouse gas emissions.

The Road Ahead

While the Earth's climate has changed naturally in the past, and will continue to do so, this fact does not absolve us of the need to address the impacts of human-induced climate change. The current rate of warming has severe implications for global ecosystems and human societies, and the window for effective action is rapidly closing.

As we move forward, the science makes it clear that we must significantly reduce greenhouse gas emissions from human activities. This requires a transition away from fossil fuels towards renewable energy sources, improvements in energy efficiency, and changes in land use practices.

The task is daunting, but there is cause for optimism. Technological advancements and economic incentives are driving a global shift towards cleaner, renewable sources of energy. More and more countries, cities, and businesses are committing to carbon neutrality, indicating a growing recognition of the need for action.

There are also steps that each of us can take in our daily lives to reduce our carbon footprint, from the foods we eat to the way we travel. Every bit counts, and collective action can lead to significant change.

The "elephant in the room", the suggestion that the current trend of climate change might simply be a natural cycle, is based on a misunderstanding of the dynamics of Earth's climate system. The reality is that the current rate of warming is unprecedented within the context of Earth's natural climate variability, and this warming trend aligns closely with the increased greenhouse gas emissions from human activities.

Natural climate cycles have shaped life on Earth for millions of years, but what we're witnessing now is different. By acknowledging our role in this change, we can take the necessary steps towards mitigating the impacts of climate change and adapting to a warmer world. The road ahead is challenging, but with concerted global effort, a sustainable and resilient future is within our grasp.

Chapter 3: Social Impacts: Community and Climate Change

Climate Refugees: An Emerging Social Class

Perhaps one of the most significant, yet underrepresented aspects of climate change, lies in its potential to displace communities, disrupt societies, and reshape human geography. Amid the flurry of melting ice caps and forest fires, the phrase "climate refugee" is gradually surfacing in global discourse.

A Definitional Dilemma

The term "climate refugee" or "climate migrant" might seem self-explanatory. However, formally defining it is an ongoing challenge. The 1951 Refugee Convention, a cornerstone of international refugee law, fails to recognize climate change-induced displacement, focusing primarily on individuals persecuted for reasons such as race, religion, or political affiliation. Consequently, individuals fleeing their homes due to climate change do not have the same legal protections that war or political refugees might receive.

Regardless of this legal conundrum, the situation on the ground is undeniable: people are being forced to leave their homes due to changes in their environment. Rising sea levels, increased frequency and intensity of natural disasters, prolonged droughts, and other climate change-related phenomena are progressively rendering parts of the planet uninhabitable.

From Bangladesh to the Pacific Islands: A Global Crisis

Climate-induced displacement is not a future hypothetical. It's happening now, and it's global. Take, for instance, the low-lying delta nation of Bangladesh. With vast sections of the country barely above sea level, even a slight rise in water levels has catastrophic implications. Flooding and riverbank erosion are already causing displacement, and as the sea level continues to rise, millions more are at risk.

In the Pacific, small island nations such as Kiribati, Tuvalu, and the Marshall Islands face existential threats from climate change. Here, the problem is two-fold: sea-level rise and storm surge. The low elevation and small landmass of these islands make them particularly vulnerable to both slowly encroaching seas and the increasingly destructive storms that accompany a warming planet.

It's not just about rising seas and cyclones, though. In sub-Saharan Africa, climate change has amplified existing challenges such as desertification and drought, driving displacement as traditional livelihoods such as farming and pastoralism become untenable.

The Social Implications

The displacement of populations due to climate change carries profound social implications. As people move, they carry with them their cultures, languages, and traditions, infusing their host societies with new influences. This process can enrich societies but can also create tensions. Competition for resources, cultural clashes, and a general sense of dislocation can lead to social friction and exacerbate existing inequalities.

Furthermore, climate-induced displacement threatens to widen the gap between the world's rich and poor. The cruel irony of climate change is that those who contributed least to the problem — often impoverished, marginalized communities — are the ones bearing the brunt of its impacts.

Towards a Solution

Tackling the challenge of climate refugees requires a multi-faceted approach. Internationally, there's a need for legal and policy frameworks that recognize climate refugees and ensure their protection. National governments, too, must prepare for internal displacement and ensure the rights and welfare of displaced populations.

Yet, at its heart, the crisis of climate refugees is a symptom of the larger problem — climate change itself. If we don't address the root causes of climate change, we are merely putting a band-aid on a bullet wound. Reducing greenhouse gas emissions and working towards global climate mitigation must remain our primary goal.

Climate refugees represent an emerging social class, a sign of the tumultuous changes that climate change is bringing to our world. As we grapple with this crisis, we must remember that the driving force behind this displacement is a global issue that affects us all. It's not simply a matter of addressing the symptoms; we must also tackle the cause.

Collaboration and Adaptation: Key Tenets of the Response

Addressing the challenge of climate refugees will require unprecedented levels of collaboration, both domestically and internationally. Climate change does not respect borders, and neither does the displacement it causes. Hence, solutions will necessitate international cooperation and shared responsibility.

Adaptation will also be key. Coastal defenses, managed retreats, and climate-resilient infrastructure can help communities withstand the effects of climate change. However, it is crucial to involve local communities in these adaptation strategies. Not only do they best understand the local context and impacts, but such involvement can also provide a sense of agency in the face of climate-induced threats.

Resilience and Empowerment: Building a More Inclusive Future

There is, however, an often-overlooked aspect to this crisis: the resilience and empowerment of the displaced communities. Labeling these communities as mere victims can overshadow their potential to actively shape their futures.

Efforts should be made to empower these communities, helping them develop the skills and resources needed to adapt to their new environments. Moreover, it's vital to recognize and respect their rights, dignity, and cultures, ensuring their voices are heard in decision-making processes.

In this era of climate change, fostering resilience is not just about building higher sea walls or drought-resistant crops. It's also about creating societies capable of absorbing shocks and stresses while still preserving the cultural and social fabric that binds us together.

Climate change is reshaping the social landscape of our world. The emergence of climate refugees presents complex challenges to our societies, our legal frameworks, and our conceptions of displacement and migration. Yet, it also offers opportunities for resilience, adaptation, and greater social inclusivity.

The issue of climate refugees is a stark reminder that climate change is not some distant, impersonal phenomenon. It is a deeply human issue, with direct and tangible impacts on individuals and communities across the globe.

If we are to navigate the age of adaptation successfully, we must not only confront the scientific and political aspects of climate change but also its social dimensions. Recognizing and addressing the plight of climate refugees is an essential step in this journey.

Climate change may be the defining challenge of our time, but it is also an opportunity - a chance for us to rethink, reimagine, and reshape our world towards one that is more sustainable, inclusive, and resilient. And perhaps, in this process, we can also reshape our minds, fostering a deeper understanding of our interconnectedness and shared responsibility for our planet and for each other.

The Role of Community: Building Resilience Against Climate Change

When we discuss the global response to climate change, the discourse often revolves around top-down, macroscopic initiatives - international climate agreements, national policies, corporate sustainability practices. Yet, the key to climate resilience might be found at the opposite end of the scale. At the community level, where everyday decisions accumulate into a collective, transformative force, lies a remarkable potential to counter the climate crisis.

The Power of Local: Understanding Community Resilience

Community resilience in the context of climate change refers to the capacity of a local community to anticipate, withstand, recover from, and adapt to climate-related shocks and stressors. This resilience is not just about infrastructure or local economies; it's also about social connections, traditions, shared knowledge, and communal action.

Climate change is a multifaceted problem that impacts different regions in unique ways. A one-size-fits-all approach is far from effective in dealing with these diverse impacts. Local communities, on the other hand, understand their unique vulnerabilities and can create bespoke resilience strategies. Whether it's a coastal community devising innovative ways to combat sea-level rise or an agricultural community switching to drought-resistant crops, community-led actions can be powerful tools in the climate resilience toolbox.

Case Studies: The Impact of Community Resilience

Across the globe, communities are proving that they can rise to the challenge of climate change. In the coastal city of Semarang, Indonesia, community groups have been leading mangrove planting efforts to curb coastal erosion and mitigate the impacts of rising sea levels.

In Africa's Sahel region, community-led Great Green Wall initiative aims to combat desertification and improve livelihoods through the creation of a green "belt" across the continent. By planting drought-resistant trees, the communities involved are not only drawing down carbon but also creating a more resilient local ecosystem and improving their quality of life.

Fostering Community Resilience: A Multi-dimensional Approach

Fostering community resilience is not merely about specific climate actions. It's about strengthening the social fabric, promoting participatory governance, and nurturing local leadership. It's about ensuring access to resources, knowledge, and decision-making processes.

Education plays a pivotal role here. Knowledge about climate change, its local impacts, and potential coping strategies empowers communities to act. Schools, community centers, local organizations can all be platforms for disseminating this knowledge and nurturing climate literacy.

It's also about creating inclusive and participatory decision-making processes. When community members feel they have a say in decisions that affect their lives, it fosters a sense of ownership and responsibility. It also ensures that the solutions devised are grounded in the realities of the community and therefore have a higher chance of being successful and sustainable.

The Challenge of Inequality

Yet, as we delve into the realm of community resilience, it's essential to address the elephant in the room: inequality. Disparities in wealth, resources, and access to power can hamper community resilience efforts. Vulnerable and marginalized groups within communities are often the most impacted by climate change but have the least resources to adapt and cope. Therefore, fostering community resilience is also about addressing these systemic inequalities. It's about ensuring that all voices within the community are heard and respected, and that resources are distributed equitably.

In the Face of Global Crisis, a Local Solution

As we navigate the turbulence of the climate crisis, communities have a crucial role to play. They are not only the first line of defense against climate impacts but also hubs of innovative, localized solutions.

By fostering community resilience, we can build a grassroots force against climate change, a force rooted in local knowledge, shared responsibility, and a deep sense of place. This process won't be easy. It will require breaking down barriers, addressing inequalities, and fostering a shared sense of purpose. It will require that we not only change our practices but also shift our mindset, recognizing the inherent value and potential of local communities.

Resilient communities are those that harness the power of unity, where each member understands their role and feels empowered to act. It involves reshaping education systems to ensure a profound understanding of climate change and its impacts. It means encouraging local businesses to adopt sustainable practices, thus creating an economy that supports and reinforces resilience.

In the face of natural disasters, resilient communities are able to protect and rebuild themselves, often becoming stronger than before. They become repositories of knowledge, experience, and practical know-how, equipping them to better respond to future crises. They become models of what can be achieved when people come together, driven by a common cause.

Technology can also play a significant role in fostering community resilience. From using geographic information system (GIS) mapping for disaster risk reduction to leveraging social media platforms for community organization and climate education, technology can equip communities with the tools needed to adapt and thrive in a changing climate.

Furthermore, community resilience initiatives can serve as powerful narratives that inspire and motivate action at other levels. They can influence policies, redirect investments, and encourage shifts in societal norms and values.

However, communities can't do it alone. They need support from the larger system – from governments, businesses, non-profits, and the public at large. This support can come in various forms, from funding and resources to supportive policies and legislation. Climate resilience must become a national and global priority, with communities at the center of these efforts.

As we stand on the precipice of climate uncertainty, community resilience offers a beacon of hope. It reminds us that in our diversity and interconnectedness lies our greatest strength. That in every community, there is an untapped potential for adaptation, innovation, and resilience.

In 'The Age of Adaptation', as we navigate the unprecedented challenges of climate change, we must turn our gaze towards this often-overlooked sphere of action. It is within our communities that we can foster the collective resilience needed to face the climate crisis, ensuring not just our survival, but also a thriving, equitable future for all.

Through community resilience, we can ensure that adaptation to climate change is not just about surviving but about thriving. It's about shaping a world that, despite the odds, can flourish in harmony with the natural world, honoring the delicate balance that sustains us all.

The age of adaptation is here, and it starts within our communities. Each one of us has a role to play in this grand narrative of resilience. And who knows? Perhaps, in the process of transforming our communities, we might just transform ourselves, reshaping our minds to fully appreciate our interdependence, our shared responsibility, and our collective power. In the face of climate change, it is together, as resilient communities, that we will forge a path towards a sustainable future.

The Impact on Culture and Identity: The Loss of "Home"

Climate change isn't just transforming our physical world; it's also altering the landscape of human culture and identity. As the seas rise, forests burn, and deserts expand, we aren't just losing homes; we're losing histories, traditions, languages - in essence, the elements that make up the rich tapestry of human diversity.

Place and Identity: The Fabric of "Home"

At the heart of this transformation is the profound connection between place and identity. 'Home' is more than just a shelter; it's a tapestry woven with threads of culture, community, history, and personal memory. It forms an integral part of our identity, grounding us in space and time, connecting us to our ancestors, and forming a bridge to our descendants.
When climate change displaces us from these places, it doesn't just displace us physically; it also dislocates us emotionally and culturally. This loss of 'home' can lead to a profound sense of rootlessness, a loss of self, and a grief that's hard to articulate - a phenomenon often referred to as solastalgia.

Displacement and Dislocation: The Climate Exodus

As sea levels rise and extreme weather events become more common, the world is witnessing a growing tide of climate refugees - people forced to leave their homes due to climate change. According to the Internal Displacement Monitoring Center, as of 2020, an average of 20 million people is displaced annually by climate-related disasters. By 2050, this number could rise to anywhere from 150 to 300 million.

However, this exodus isn't just about numbers; it's about the human stories behind these figures. Each displaced individual or community carries with them a unique culture, a unique identity rooted in a specific place. When they are forced to leave, part of this identity remains behind - in the landscapes, the traditions, the languages that are intimately tied to their original homes.

Culture on the Frontlines: The Case of Indigenous Communities

Indigenous communities, who have been the guardians of their lands for centuries, are at the forefront of this cultural loss. Their identities are intrinsically tied to their ancestral lands, their cultures shaped by the rhythms of the seasons, the patterns of the wildlife, the contours of the landscapes. From the Inuit communities in the Arctic, witnessing the melting ice caps, to the Pacific Islanders, grappling with rising sea levels, climate change is an existential threat to these cultures. As their habitats transform, they're not just losing their homes; they're also losing their languages, their rituals, their collective memory - the very essence of their cultural identity.

The Loss of Cultural Diversity: A Silent Crisis

The loss of cultural diversity that accompanies the climate crisis is a silent crisis. Unlike the physical impacts of climate change, this loss is often invisible, intangible. However, this invisibility doesn't make the loss any less significant. The erosion of cultural diversity is akin to the erosion of biodiversity - it makes our world less resilient, less vibrant, less capable of adapting to new challenges.

Cultural diversity is humanity's heritage, our shared wealth. It enriches our understanding of the world and ourselves. It broadens our perspectives, enables us to learn from each other, and fosters innovation and adaptability. As we lose cultures, we lose these invaluable resources. We lose pieces of our shared human story.

Grief, Mourning, and Healing: Acknowledging the Loss

Addressing the cultural loss wrought by climate change begins by acknowledging it. By recognizing the grief, the mourning that accompanies the loss of 'home.' This recognition validates the experiences of those most impacted and creates a space for healing.

Storytelling can play a crucial role in this process. Stories have the power to make the invisible visible, to give voice to the voiceless. By telling the stories of those impacted by climate change, we can bring the cultural loss into the collective consciousness. These stories can deepen our understanding of climate impacts, humanize the statistics, and foster empathy and solidarity.

Therapeutic approaches such as eco-therapy and climate counseling can also provide much-needed support for those grappling with the psychological impacts of climate change. These therapies can help individuals and communities process their loss, manage their eco-anxiety, and find ways to adapt and build resilience in the face of change.

Cultural Adaptation and Resilience: Charting a New Course

While the loss of 'home' is a profound challenge, it's not the end of the story. Cultures, like ecosystems, are dynamic. They change, adapt, evolve. Faced with the climate crisis, many communities are finding innovative ways to preserve their cultural identities, even as they navigate the shifting landscapes of their physical homes.

From creating digital archives of disappearing languages and traditions to developing climate-resilient practices that align with cultural values, these communities are charting a new course for cultural preservation in the age of climate change. They are redefining 'home' in a world reshaped by the climate crisis.

The Role of Global Community: Towards Cultural Sustainability

Just as the climate crisis requires a global response, so too does the cultural crisis it engenders. The loss of cultural diversity is a loss for all of humanity, and it calls for a collective response. Cultural sustainability should be an integral part of climate action, recognizing that cultural diversity is as crucial to our survival and wellbeing as biodiversity.

This approach would involve legal protections for climate-displaced communities, policies that acknowledge and address the cultural impacts of climate change, and measures to support the cultural adaptation and resilience of these communities. It would also involve amplifying the voices of these communities in global dialogues on climate change, ensuring their experiences and perspectives are heard and valued.

Home in the Age of Adaptation

As we navigate the age of adaptation, we must not forget that 'home' is more than just a physical place. It's the ground beneath our feet, the stories we tell, the rituals we perform, the languages we speak. It's the thread that connects us to our past, grounds us in the present, and reaches out to our future.

Climate change is not just transforming our physical homes; it's also transforming our cultural landscapes, reshaping the very concept of 'home.' In the face of this change, our challenge is to find ways to preserve the essence of our cultural identities, even as we adapt to new realities.

In the age of adaptation, 'home' may look different. But it can still provide us with a sense of belonging, a sense of continuity, a sense of identity. Even in a changing world, 'home' can still be a source of strength, resilience, and hope. It's up to us to ensure that it is.

Chapter 4: The Politics of Climate Change: A Global Challenge

Climate Change Policy: Global Winners and Losers

In our interconnected world, climate change is as much a political issue as it is an environmental one. The implications of global warming have become a major talking point in policy discussions, shaping diplomatic relations, and driving economic strategies. From boardrooms to war rooms, from local communities to international platforms, the conversation around climate change has become pivotal, revealing a new global order of winners and losers.

Understanding The Politics of Climate Change

At the core of the political debate on climate change is the issue of responsibility and capacity. While the historical responsibility for climate change largely rests with industrialized nations, the countries most vulnerable to its impacts are often those with the least financial and technological means to combat it. This imbalance has created a complex tapestry of geopolitics, where international collaboration becomes a daunting challenge.

Understanding the politics of climate change requires acknowledging this imbalance. It also involves recognizing the potential opportunities that arise from transitioning towards a low-carbon economy. As nations grapple with the task of mitigating and adapting to climate change, new power dynamics emerge, creating a new breed of winners and losers.

The Winners: Opportunities in The Green Economy

In the green transition, nations investing heavily in renewable energy, green technologies, and climate-resilient infrastructures stand to gain significantly. These 'winners' recognize the economic opportunities that come with the shift towards a sustainable, low-carbon economy. Countries like Denmark, Germany, and China are prime examples. They've invested extensively in renewable energy and green technologies, transforming themselves into global leaders in the green economy. This shift doesn't only reduce their carbon emissions, but it also bolsters their economies, creating jobs, fostering innovation, and improving energy security.

Emerging economies also have an opportunity to leapfrog the fossil fuel-based development model and embrace green growth. Nations such as India and Kenya are increasingly investing in renewable energy, skipping the traditional development pathway and moving directly to a more sustainable model.

The Losers: The High Cost of Inaction

On the other hand, countries that fail to act on climate change or are slow to adapt face considerable risks. Economies heavily reliant on fossil fuels, such as Russia and Saudi Arabia, stand to lose as the world moves towards renewable energy. The declining demand for fossil fuels could leave these nations with stranded assets, potentially leading to significant economic loss.

Vulnerable nations, such as the small island states, face the brunt of climate impacts despite their minimal contribution to the problem. Rising sea levels, stronger hurricanes, and saline intrusion threaten their very existence, yet they often lack the necessary resources to effectively adapt to these changes.

Climate Justice: Bridging the Gap

In the grand chessboard of climate politics, climate justice emerges as a crucial strategy to bridge the gap between the winners and the losers. It advocates for fairness in the allocation of burdens and benefits related to climate change.

Climate justice emphasizes that the nations and communities that have contributed least to climate change - who are often the most vulnerable to its impacts - should not bear the brunt of its consequences. It argues for greater financial and technological support from developed countries to developing nations, to help them adapt to climate impacts and transition towards a low-carbon economy.

The Politics of Adaptation: The Path Ahead

The politics of climate change is intricate and evolving. As nations navigate their way through the challenges and opportunities of climate change, international cooperation remains key. The 2015 Paris Agreement, while not perfect, represents a global consensus to tackle the climate crisis. Its focus on a fair and equitable transition to a low-carbon future embodies the principle of climate justice.

However, it is evident that rhetoric needs to be matched with tangible action. To mitigate the impacts of climate change and adapt to its reality, nations need to drastically ramp up their climate ambitions. This includes enhancing their Nationally Determined Contributions (NDCs), investing more in renewable energy and climate-resilient infrastructure, and phasing out fossil fuel subsidies.

Furthermore, climate finance remains a significant piece of the puzzle. Developed nations have pledged to provide $100 billion per year by 2020 to support developing countries in their climate actions. However, this goal is yet to be reached. Ensuring that these funds are adequately provided and effectively utilized is crucial for global climate action.

Climate diplomacy also plays a vital role. The international climate negotiations under the United Nations Framework Convention on Climate Change (UNFCCC) offer a platform for nations to collaborate on climate action. Ensuring that these negotiations are inclusive, transparent, and effective is essential for driving global climate action.

The Human Dimension: Beyond Politics

While the politics of climate change often focuses on nations, we must remember that climate change is ultimately about people. It's about the farmer in Bangladesh grappling with unpredictable monsoons, the Indigenous community in the Arctic witnessing the melting ice, the city dweller in Beijing choking on smog, and the young activist in New York marching for her future.

Climate action needs to be people-centered. This involves not only protecting people from the impacts of climate change but also ensuring that they are active participants in the decision-making processes. It involves recognizing and respecting the rights of Indigenous peoples and local communities, who are often the frontline defenders of nature. It also involves empowering the youth, the inheritors of our planet, to shape their future.

Conclusion: Towards A Resilient and Just Future

Climate change is a complex and multifaceted challenge, and its political dimensions are no exception. It has the power to disrupt the existing global order, creating new winners and losers. However, it also offers an opportunity to reshape this order, to build a more resilient and just world.

In this world, the winners wouldn't be determined merely by their economic power, but by their ability to foster sustainable growth, to build inclusive and resilient societies, and to uphold the principles of climate justice. The losers, on the other hand, would be those who choose short-term gains over long-term sustainability, who cling to the old ways while the world moves forward.

As we navigate the age of adaptation, our challenge is to ensure that this new world is not just resilient to the physical impacts of climate change, but also to the political and social upheavals it entails. This requires a fundamental shift in our mindset, from seeing climate change as a distant environmental problem to recognizing it as an immediate social, economic, and political challenge.

In the end, the true winners in the age of adaptation would be those who understand this challenge, who choose cooperation over conflict, justice over inequity, and sustainability over short-termism. They would be those who choose not just to survive in a changing world, but to thrive.

The Emergence of Climate Activism and Its Impact on Policy

In an age characterized by daunting climate threats and heightened political divisiveness, an unexpected ray of hope has emerged from the grassroots. Climate activism, led by concerned citizens, environmental groups, and a bold, outspoken generation of young people, has made its way from the fringes of public discourse to the mainstream political arena.

In this section, we will trace the rise of climate activism, exploring its origins, its various forms, and most crucially, its profound impact on climate policy.

A Brief History: Origins of Climate Activism

The roots of climate activism extend back to the broader environmental movement that gained momentum in the mid-20th century. Early activists like Rachel Carson, whose book "Silent Spring" sparked an awakening about the ecological impacts of industrial chemicals, laid the groundwork for public awareness about human impacts on the environment.

However, it wasn't until the 1980s, with increasing scientific understanding of global warming, that climate-specific activism began to emerge. Organizations such as Greenpeace and Friends of the Earth started focusing on climate change, launching campaigns to raise awareness and demand policy changes.

A New Wave: The Rise of Youth Activism

The turn of the millennium saw an acceleration of climate activism, with a marked shift in its demographic makeup. The new generation of climate activists were younger, bolder, and more globally connected.

The youth-driven climate strikes, popularly known as 'Fridays for Future,' exemplify this trend. Sparked by Swedish teenager Greta Thunberg's solitary protest in 2018, the movement has since proliferated across the globe, mobilizing millions of young people demanding urgent climate action.

This new wave of activism differs from its predecessors in its sense of urgency, scale, and inclusivity. Young activists, acutely aware of the existential threat posed by climate change, are not just pleading for action but demanding it. They've harnessed the power of social media, transcending geographical and cultural barriers, to mobilize on an unprecedented scale. They've also championed inclusivity, ensuring voices from the Global South, indigenous communities, and marginalized groups are heard.

Impacts on Policy: From Streets to Parliaments

One might wonder, amidst the chants and placards, the strikes and marches, is climate activism effecting real change? The answer, while complex, is increasingly positive. While the pathways from protest to policy can be indirect and convoluted, several concrete examples illustrate the influence of climate activism on policy-making.

In many countries, activists have used the courts to advance their cause. Landmark lawsuits have been filed against governments and corporations for failing to address climate change. One notable example is the 2015 Urgenda case in the Netherlands, where the government was legally obliged to cut greenhouse gas emissions significantly.

Policy changes can also be traced back to direct activist pressure. For instance, the Green New Deal, a comprehensive policy proposal in the United States aiming to address climate change and economic inequality, gained traction largely due to the advocacy by the youth-led Sunrise Movement.

Moreover, climate activism has had considerable success in shifting public opinion. A growing body of research shows that mass protests not only raise awareness but also influence public attitudes and beliefs about climate change. This shift in public sentiment can, in turn, lead to increased political prioritization of climate change.

Challenges and Future Trajectories

While climate activism has achieved significant milestones, it also faces numerous challenges. Activists often grapple with political pushback, public apathy, or even outright hostility. Moreover, while they've successfully amplified the urgency of climate change, translating this into substantial policy action remains an uphill battle.

Despite these challenges, the future of climate activism looks promising. With increasing public support, growing media coverage, and expanding alliances across various societal sectors, the movement seems poised for continued influence. As the impacts of climate change become ever more palpable, the call to action echoed by activists worldwide will likely grow louder.

Looking forward, we can anticipate several trends in climate activism. First, the movement will continue to diversify. What started as primarily an environmental cause has expanded to intersect with issues of social justice, indigenous rights, and economic inequality. This intersectionality broadens the movement's appeal and integrates climate action into wider societal transformations.

Second, activists will increasingly utilize legal avenues to demand action. The concept of 'climate justice' has taken root, arguing that climate change is not only an environmental crisis but also a legal and ethical one. This perspective opens up new possibilities for holding governments and corporations accountable.

Lastly, we'll likely see an increase in 'constructive' activism. While protesting and campaigning against harmful practices are vital, there's a growing focus on advocating for positive alternatives. This involves engaging with political processes, collaborating with sympathetic businesses, and promoting sustainable community initiatives.

The emergence of climate activism constitutes a transformative force in the global response to climate change. From its humble beginnings, the movement has evolved into a dynamic, influential, and globally connected network of citizens demanding change. Young activists, in particular, have reinvigorated the cause, bringing a sense of urgency and inclusivity that has shifted public attitudes and influenced policy discussions.

Although challenges abound, the trajectory of climate activism provides reasons for optimism. With its adaptive strategies, diverse alliances, and unwavering commitment, the movement has proven its capacity to shape public discourse and policy decisions alike. As we chart our course through the age of adaptation, the role of climate activism as a catalyst for change seems not only important but indispensable.

In the final analysis, climate activism represents a compelling demonstration of our collective capacity to face the climate crisis head-on. It exemplifies the power of human agency in shaping our future, reminding us that even as we are agents of climate change, we are also agents of climate action. Amid the grim projections and daunting challenges, it is a testament to our potential to influence, innovate, and inspire – a beacon of hope in the age of adaptation.

We must remember that, ultimately, climate change is not just a technical issue to be solved by science or a political issue to be negotiated by governments. It is a human issue, one that calls for a collective response from all of us. As the climate activists remind us, we are all part of the solution.

The challenge of climate change, then, is not simply about adapting to a changing world. It's about changing ourselves. It's about changing our behaviors, our policies, our societies. It's about changing our world, for the better.

The Power of Green Diplomacy: Climate Change and International Relations

Our world is an interdependent one. The invisible threads of commerce, technology, culture, and ecology connect us in intricate patterns of interaction. Nowhere is this more evident than in the field of climate change – an issue that, like no other, transcends borders and jurisdictions, affecting every country and every individual, albeit in disparate ways.

In this chapter, we explore the pivotal role of green diplomacy in international relations amidst our changing climate. "Green diplomacy" refers to diplomatic efforts aimed at addressing environmental issues, particularly climate change. In the modern geopolitical landscape, these efforts are shaping international politics and influencing the global power dynamics.

The Emergence of Green Diplomacy

The dawn of the 21st century has seen environmental issues become central in the international diplomatic arena. Global warming, deforestation, ocean acidification, and species extinction are now commonplace topics at the United Nations (UN), G20 meetings, and even military strategic talks.

The Paris Agreement of 2015 was a watershed moment for green diplomacy. For the first time, almost all countries pledged to limit global warming to well below 2 degrees Celsius and to strive for 1.5 degrees. This ambitious treaty represented a seismic shift in the diplomatic landscape, indicating that environmental stewardship had moved from the periphery to the core of international relations.

However, despite these commitments, implementing and achieving the Paris Agreement targets has been an uphill battle. Green diplomacy, therefore, has shifted its focus from making commitments to ensuring accountability and action.

Green Diplomacy and Global Power Dynamics

Green diplomacy is increasingly influencing global power dynamics. Traditional powerhouses like the United States, the European Union, and China find their climate policies scrutinized and used as a gauge of their global leadership. At the same time, smaller nations, especially those most vulnerable to climate impacts, are finding a powerful voice on the international stage through climate advocacy.

Moreover, countries are beginning to use green diplomacy as a strategic tool, recognizing that climate leadership can translate into enhanced geopolitical influence. For instance, the European Union has leveraged its ambitious Green Deal as a diplomatic tool, aiming to set a global benchmark and export its climate standards worldwide.

In contrast, countries failing to tackle climate change or reneging on their commitments may find themselves ostracized. The temporary withdrawal of the United States from the Paris Agreement under President Trump was met with global criticism and served to isolate the country on the international stage.

Furthermore, we are witnessing the evolution of a new kind of diplomacy, known as "climate diplomacy." This approach integrates climate change into all aspects of foreign policy, viewing it as a critical component of peace and security, economic development, and human rights. In a climate-constrained world, a country's stance on climate change is now as significant as its military might or economic prowess.

Challenges and Opportunities in Green Diplomacy

Green diplomacy is fraught with challenges. There are stark inequalities in countries' contributions to climate change and their capacities to deal with its impacts. Balancing these inequalities while striving for ambitious global action is a delicate diplomatic dance.

Moreover, climate change is a "super wicked problem," characterized by deep uncertainties, long time horizons, and high stakes. These attributes make it a uniquely challenging issue to address through traditional diplomatic avenues.

Despite these challenges, green diplomacy also presents vast opportunities. It fosters international cooperation in a time of increasing geopolitical tension. It enables the sharing of innovative solutions, from renewable energy technologies to strategies for climate adaptation. It also promotes global equity by giving a voice to vulnerable countries and populations disproportionately affected by climate change.

Looking Ahead: The Future of Green Diplomacy

As we move deeper into the 21st century, green diplomacy will continue to reshape the international landscape. As the impacts of climate change become more pronounced, environmental issues will no longer be just another item on the diplomatic agenda but a key driver of international politics.

The role of climate change in sparking conflict or exacerbating existing tensions will necessitate more integrative, climate-conscious approaches to peacekeeping and conflict resolution. Nations will need to reassess alliances and rivalries in light of shared climate vulnerabilities and opportunities. Climate change will increasingly be framed not just as an environmental issue, but as an existential threat to national security, human rights, and social stability.

Moreover, green diplomacy is likely to spur the creation of new diplomatic structures and institutions. Existing bodies like the United Nations Framework Convention on Climate Change (UNFCCC) will be strengthened, and we might see the formation of new, specialized forums for climate negotiation and cooperation. International law may also evolve to better handle transboundary climate issues, such as climate-induced migration or the attribution of extreme weather events to specific countries' emissions.

Green diplomacy also holds the potential to rebalance global power structures. Emerging economies with ambitious climate action plans could carve out influential roles on the world stage. Small Island Developing States

(SIDS) and other vulnerable countries could also wield greater influence as the urgency of their plight impels larger nations to act.

Green diplomacy might also inspire new forms of multilateralism. Climate change is a collective problem that requires collective solutions. Traditional 'state-centric' diplomacy may give way to more inclusive forms that engage non-state actors, including cities, businesses, and civil society groups. This 'multi-stakeholder' diplomacy recognizes that addressing climate change requires an all-hands-on-deck approach.

Finally, as green diplomacy becomes increasingly prominent, we are likely to see greater accountability in international climate politics. Diplomatic pressure, public scrutiny, and legal action will be employed to hold countries accountable for their climate commitments. This will be accompanied by a shift towards transparency and robust reporting, with countries expected to provide clear, credible evidence of their progress towards climate goals.

As we navigate through the uncharted waters of a changing climate, green diplomacy will be our compass. It will guide our collective efforts to mitigate greenhouse gas emissions, adapt to unavoidable changes, and create a more sustainable and equitable world.

The power of green diplomacy is not just in its capacity to influence policies or negotiate treaties. It's in its ability to foster understanding, build consensus, and spur collective action. It is the instrument through which we can transform the challenge of climate change into an opportunity for global cooperation and sustainable development.

As such, green diplomacy represents the marriage of two of humanity's greatest achievements: our ability to create cooperative societies, and our ability to understand and shape the world around us. It is a testament to our capacity to face global challenges head-on, to learn, adapt, and ultimately, to thrive.

By harnessing the power of green diplomacy, we can ensure that the legacy we leave for future generations is not one of conflict and degradation, but of resilience, innovation, and shared prosperity. Climate change may be reshaping our world, but through green diplomacy, we can shape our response.

Chapter 5: The Economics of Climate Change: Risks and Opportunities

Climate Change and Economic Inequality: Widening the Gap?

Climate change and economic inequality are two of the most pressing issues of our time. While they may seem distinct, they are intimately interconnected. Climate change amplifies and exacerbates existing economic disparities, while economic inequality can hinder effective responses to climate change. This interplay between environmental and economic factors is complex and multi-faceted, and its understanding is crucial to developing just and sustainable solutions.

The Economics of Climate Vulnerability

Climate change affects everyone, but it does not affect everyone equally. The poor are disproportionately vulnerable to the impacts of climate change, often because they have fewer resources to adapt and recover.

Let's take the example of extreme weather events, such as hurricanes, floods, and heatwaves, which are becoming more frequent and severe due to climate change. These disasters can wipe out homes, livelihoods, and infrastructure, and recovering from them can require significant resources. For individuals and communities living in poverty, these events can be catastrophic. They often lack the resources to rebuild or relocate, which can trap them in a cycle of poverty and vulnerability.

But the injustice doesn't end there. Low-income communities and countries also tend to contribute the least to greenhouse gas emissions—the primary driver of climate change. Yet, they bear the brunt of the impacts. This imbalance between contribution to the problem and the suffering from its effects is a stark example of climate injustice.

The Economic Costs of Climate Change

Climate change is not just a moral issue; it's an economic one as well. The financial costs of climate change are enormous and growing. These include direct costs, such as damage to infrastructure from extreme weather events, and indirect costs, like the lost productivity due to heat-related illnesses.

These economic losses are likely to be unevenly distributed. Countries and regions that are already economically disadvantaged are likely to bear a disproportionate share of these costs. This is not only because of their greater vulnerability to climate impacts, but also because they may have less capacity to invest in adaptation and resilience measures.

For example, consider the impact of sea-level rise on small island developing states. Many of these nations face existential threats from climate change, with entire communities at risk of being submerged. The cost of protecting or relocating these communities is enormous, and for many of these nations, it represents a significant proportion of their GDP.

Climate Change, Economic Inequality, and the Vicious Cycle

The relationship between climate change and economic inequality is not one-way. They reinforce each other in a vicious cycle. Climate change exacerbates economic inequality, and in turn, economic inequality hinders our ability to tackle climate change.

Economic inequality can create political and social barriers to climate action. Wealthy, powerful groups may resist changes that threaten their interests. For example, fossil fuel industries have historically lobbied against climate regulations. In countries where economic power is highly concentrated, such resistance can be particularly effective.

The Silver Lining: Opportunities for Climate Justice

Despite these grim realities, there are also reasons for hope. There are numerous ways in which climate action can help reduce economic inequality.

For one, the transition to a low-carbon economy can create new jobs and industries. Renewable energy, energy efficiency, and other 'green' sectors can offer economic opportunities for individuals and communities left behind by the fossil fuel economy.

Climate policies can also be designed to be progressive, benefiting lower-income households. For instance, revenue from carbon pricing can be returned to citizens as a dividend, which can help offset any increased costs of energy or goods.

Moreover, efforts to adapt to climate change can also help reduce poverty and inequality. For example, investments in climate-resilient infrastructure can protect communities from extreme weather events, while also stimulating economic activity.

Addressing climate change and economic inequality together is not just a matter of justice—it's also a matter of necessity. If we fail to address economic inequality, we risk undermining our efforts to combat climate change. Conversely, if we fail to address climate change, we risk exacerbating economic disparities.

In this age of climate change, we must rethink our economic structures. The current system, with its stark inequalities and unsustainable practices, is clearly untenable. Instead, we must strive for a more equitable, resilient, and sustainable economy.

There is a growing recognition of the need for 'just transition'—a shift to a low-carbon economy that is fair and inclusive. This concept encompasses various dimensions, including ensuring quality jobs in green industries, protecting workers in declining fossil fuel sectors, and addressing the needs of communities that are most vulnerable to climate impacts.

One example of this approach in action is the Green New Deal, a proposal that has gained traction in the United States and elsewhere. The Green New Deal seeks to address climate change and economic inequality simultaneously, by investing in green jobs, infrastructure, and social programs.

At the heart of this idea is the recognition that the fight against climate change is not just about reducing emissions—it's also about building a better, fairer world.

Another economic opportunity lies in the field of climate finance. The vast resources required to mitigate and adapt to climate change present an opportunity for innovative financial solutions. Green bonds, climate-aligned loans, and impact investing are all tools that can mobilize private capital for climate action. These financial instruments not only deliver environmental benefits but also have the potential to generate substantial economic returns.

It is also important to recognize that economic growth and carbon emissions do not have to be inherently linked. The concept of 'decoupling' suggests that it is possible to grow our economies while reducing greenhouse gas emissions. Many developed countries have already shown signs of decoupling, and the challenge is to accelerate this trend globally, particularly in fast-growing emerging economies. Moreover, investing in climate change adaptation and resilience can also have significant economic benefits. A report from the Global Commission on Adaptation found that investing $1.8 trillion globally in five areas (early warning systems, climate-resilient infrastructure, improved dryland agriculture, mangrove protection, and water resources resilience) from 2020 to 2030 could generate $7.1 trillion in total net benefits.

Climate change and economic inequality are two sides of the same coin. We cannot address one without addressing the other. But with thoughtful, ambitious policies, we have the opportunity to tackle these twin challenges together. The path forward is not easy, but it is one that can lead us to a more equitable and sustainable world. In this age of adaptation, we have not only the responsibility, but also the opportunity, to reshape our world—and our economies—for the better.

Green Economy: The Race for Sustainable Development

As our world becomes increasingly aware of the impending reality of climate change, a significant shift is taking place across the globe. This shift is moving us away from an economy reliant on fossil fuels and non-renewable resources towards one that values and invests in sustainable and renewable resources. This transition, often referred to as the "Green Economy," is not simply a trend, but a necessary transformation that represents our best chance at mitigating the most severe impacts of climate change.

So, what exactly is a Green Economy? In essence, it's an economic system that aims for sustainable development without degrading the environment. It's about reducing our carbon footprint, promoting social inclusion, and creating jobs that contribute to preserving or restoring the quality of the environment. From agriculture, manufacturing, construction, and power generation, to infrastructure, transportation, waste management, and tourism, the Green Economy encompasses all sectors and could bring about substantial economic, social, and environmental benefits.

This 'race' for sustainable development is powered by an increasing understanding that the traditional model of development—often characterized by unchecked consumption and environmental negligence—is not only harmful for our planet but also economically short-sighted. A green economy, in contrast, recognizes that economic growth and environmental responsibility are not mutually exclusive, but can, and should, go hand in hand.

The renewable energy sector is one of the most prominent examples of the Green Economy in action. Over the past few years, we've seen an unprecedented boom in renewable energy technologies, primarily wind and solar power. The International Renewable Energy Agency (IRENA) reported that by the end of 2022, the global renewable generation capacity increased to 2795 Gigawatts (GW), with solar energy and wind energy accounting for 90% of all new power capacity. This shift is significant not just for reducing greenhouse gas emissions, but also for job creation. As per IRENA's data, renewable energy jobs reached 12.5 million globally in 2022.

On a larger scale, governments and businesses are investing in green infrastructure. This includes projects like building energy-efficient buildings, improving water and waste management systems, and developing public transportation to reduce carbon emissions. The European Union, for example, has pledged to become carbon neutral by 2050 under its European Green Deal. To achieve this, they have outlined various strategies and actions, including investing significantly in green infrastructure and technologies.

Another critical component of the green economy is the circular economy model. Instead of the traditional 'take-make-waste' linear model, a circular economy emphasizes 'reduce, reuse, recycle.' It aims to design out waste, keep products and materials in use, and regenerate natural systems. This circular approach can bring substantial economic benefits. The Ellen MacArthur Foundation estimates that the shift to circularity could add $4.5 trillion to the global economy by 2030.

Similarly, green finance is gaining momentum. Green bonds, a type of fixed-income instrument dedicated to raising money for climate and environmental projects, are becoming increasingly popular. According to the Climate Bonds Initiative, green bond and green loan issuance reached $269.5 billion globally in 2022, a 45% increase from the previous year.

Meanwhile, ESG (Environmental, Social, and Governance) investing, which involves considering environmental, social, and governance factors in investment decisions, has seen dramatic growth. As of 2023, ESG assets are on track to hit $53 trillion, representing more than a third of all global assets under management, according to Bloomberg Intelligence.

However, the transition to a green economy is not without challenges. It requires significant policy reforms, technological innovation, and financial investments. Also, the green transition must be just and inclusive. It's crucial to ensure that the benefits of the green economy are shared equitably, and that workers and communities affected by the transition away from fossil fuels are not left behind. This concept, often referred to as a 'Just Transition,' is an essential component of the Green Economy.

Moreover, the greening of industries often requires reskilling of the workforce. In regions heavily dependent on industries like coal mining or oil extraction, job losses are a real concern. Therefore, governments and corporations must invest in education and training programs to prepare workers for new, green jobs. There are positive signs in this regard, with initiatives such as the National Skills Coalition in the U.S., working towards a more inclusive green economy by advocating for training and career pathways in sustainable industries.

The move towards green economies is also revealing new frontiers in the technological landscape. Carbon capture and storage (CCS) technology, energy storage solutions, and green hydrogen are just a few of the innovations that have the potential to revolutionize our approach to energy and industry. Companies such as Climeworks in Switzerland are leading the way with direct air capture technology, while others are pioneering in the field of hydrogen fuel cells. These developments underline that the green economy is as much a technological revolution as an economic one.

There's also a growing acknowledgment of the role of nature-based solutions – such as forest conservation, sustainable agriculture, and wetland restoration – in the green economy. These solutions not only help to reduce CO_2 emissions but also create jobs and support biodiversity. According to the World Economic Forum, transitioning to a nature-positive economy could create 395 million jobs by 2030.

The 'race' for sustainable development is not without its hurdles. For instance, developing countries may lack the necessary resources to invest in green technologies and infrastructure. That's where international cooperation and climate finance play a pivotal role. Financial mechanisms such as the Green Climate Fund aim to assist developing countries in adaptation and mitigation practices to counter climate change. However, climate finance needs to be significantly scaled up to meet the immense need.

The green economy presents a path forward that respects our planetary boundaries while also offering substantial economic opportunities. Its potential to create jobs, stimulate innovation, and reduce inequality makes it an essential element in our global response to climate change. Yet, the transition requires concerted effort and cooperation from all segments of society—governments, businesses, and individuals. And as we navigate this path, we must keep in mind that this race is not a competition among nations, but a collective effort for the survival and prosperity of our shared home, planet Earth.

The green economy is our future, but its success hinges on the choices we make today. The challenges are monumental, but so are the opportunities. As the famous saying goes, "We are the first generation to feel the effect of climate change and the last generation who can do something about it." Let's make sure we rise to the occasion.

Financial Systems and Climate Change: From Risk to Opportunity

It's no longer business as usual in the financial sector. The realities of climate change have swept across the financial landscape, inducing a sea-change in the way banks, investors, and insurance companies operate. In Chapter 5: The Economics of Climate Change: Risks and Opportunities, we've explored how climate change influences economic inequality and the budding promise of a green economy. Now, let's delve into how financial systems are adjusting to this new climate reality, recognizing both the risks and opportunities that climate change presents.

We'll start with the risks, as they were the initial spark that ignited the financial world's attention to climate change.

The term "physical risks" has taken on a whole new meaning. It now refers to the direct damages caused by extreme weather events and long-term changes in climate patterns. The escalating number and intensity of wildfires, storms, and floods, along with sea-level rise and shifts in agricultural productivity, all pose significant threats to the assets financial institutions invest in or insure.

A case in point is the devastating California wildfires in 2018, which led to the bankruptcy of the state's largest utility company, Pacific Gas & Electric. With over $50 billion in assets, the bankruptcy was a stern reminder for investors and lenders of the tangible risks that climate change could pose to their portfolios.

Then, we have "transition risks" — financial risks that could result from the transition to a low-carbon economy. If the shift happens abruptly or is inadequately managed, companies heavily reliant on fossil fuels could see their assets become stranded and lose value rapidly. This potential devaluation is not a distant possibility but is already happening. In 2020, BP slashed the value of its assets by $17.5 billion, acknowledging that some of its oil and gas reserves might never be extracted due to declining demand and increasing regulation.

Yet, every cloud has a silver lining. Alongside risks, climate change also brings opportunities, and forward-looking financial institutions are keen to capitalize on them.

As governments and businesses align with the Paris Agreement's goal to limit global warming to well below 2 degrees Celsius above pre-industrial levels, there is a growing demand for "green finance." This includes investments in renewable energy, energy-efficient infrastructure, and other projects that help reduce greenhouse gas emissions or adapt to climate change impacts.

The green bond market is a shining example. These bonds finance projects with environmental benefits, and the market has skyrocketed from virtually zero in 2007 to an estimated $1 trillion by the end of 2021.

Likewise, the rise of Environmental, Social, and Governance (ESG) investing, where climate considerations are a major factor, has been nothing short of remarkable. As of 2021, ESG assets were on track to exceed $53 trillion by 2025, making up a third of global assets under management.

Another area ripe with opportunity is climate risk assessment and management. In a world increasingly threatened by climate change, tools that help investors assess and manage climate-related risks are becoming invaluable. Companies specializing in climate risk analytics, like BlackRock's Aladdin Climate, are growing rapidly, signifying a lucrative market.

However, the journey from risk to opportunity is not an easy one. It requires a comprehensive shift in how financial systems operate — one that is already underway.

Policymakers are stepping up, recognizing the need for a robust regulatory framework. Central banks and financial regulators worldwide, under the umbrella of the Network for Greening the Financial System (NGFS), are working to ensure the financial system's stability in the face of climate change. This includes efforts to incorporate climate risks into stress testing and to improve the disclosure of climate-related risks.

The Task Force on Climate-related Financial Disclosures (TCFD), established by the Financial Stability Board, has led the charge in enhancing transparency. Since its inception in 2015, the TCFD has proposed a set of recommendations that have now been adopted by over 2,000 organizations worldwide, representing a market capitalization of over $20 trillion. These recommendations, which focus on governance, strategy, risk management, and metrics and targets, are fostering a culture of openness that aids investors in making informed decisions.

However, there is room for growth in data quality, comparability, and consistency. Multiple reporting frameworks and the lack of standardized climate risk metrics can lead to information overload and make assessments challenging. Hence, efforts are underway to streamline the reporting landscape and develop consistent climate risk measures.

Another key challenge is the integration of climate risks into day-to-day decision-making. While there's growing recognition of these risks, many financial institutions still struggle to quantify them and integrate them into their risk assessment processes. Advances in climate science and technology are helping to bridge this gap. For instance, initiatives like the Partnership for Carbon Accounting Financials (PCAF) are providing methodologies for financial institutions to measure and disclose their greenhouse gas emissions.

Climate change is also opening new avenues for innovation and collaboration. Fintech — the integration of technology into financial services — is playing a crucial role in navigating the climate crisis. From peer-to-peer lending platforms that support green businesses to digital platforms that allow individuals to invest in climate solutions, fintech is democratizing access to green finance. Meanwhile, collaboration among different stakeholders — governments, financial institutions, businesses, and civil society — is proving instrumental in driving the financial system's green transformation. A notable example is the Glasgow Financial Alliance for Net Zero (GFANZ), a coalition of over 450 firms (as of 2023) committed to accelerating the decarbonization of the economy and limiting global warming to 1.5 degrees Celsius.

Despite the considerable strides made, we're still in the early stages of understanding and responding to the multifaceted implications of climate change for our financial systems. Greater efforts are needed to ensure these systems are resilient and can support the transition to a low-carbon, climate-resilient economy.

As we move forward, there's a need to keep sight of the overarching goal: not just to survive in this new climate reality, but to leverage it as an opportunity to build a more sustainable, equitable, and resilient global economy. The risks are significant, but so too are the opportunities. In this Age of Adaptation, it's upon us to rise to the challenge and turn the tide on climate change.

It's clear that our financial systems are at a crossroads, caught between the rock of risk and the hard place of opportunity. The way ahead is uncertain and fraught with challenges, but it also offers the promise of a more sustainable, resilient, and equitable future.

As we venture into uncharted territory, our choices and actions will shape not only the future of finance but also the future of our planet. And that's a responsibility we can't afford to take lightly.

Chapter 6: A Paradigm Shift: Embracing Climate Positivity

The Concept of Climate Positivity: Beyond Sustainability

In the age of climate change, concepts like 'sustainability' and 'resilience' have often taken center stage in conversations around environmental stewardship. These terms speak to the desire for endurance and the ability to withstand the shocks and disturbances that our planet is increasingly experiencing. But as we delve deeper into the Anthropocene, it's becoming evident that simply enduring is not enough. We need to aspire for more. This is where the concept of 'Climate Positivity' comes into play.

The term 'Climate Positivity' or 'Climate Positive' is relatively new to the environmental lexicon. It emerged as a more ambitious successor to the notion of 'Carbon Neutrality.' While the goal of carbon neutrality is to balance the amount of carbon dioxide we release into the atmosphere with the amount we remove, becoming climate positive means going a step further. A climate positive action or organization not only neutralizes its carbon footprint but also creates an environmental benefit by removing additional carbon dioxide from the atmosphere.

But climate positivity goes beyond just carbon sequestration. It extends to every facet of our relationship with the Earth, from how we generate energy and produce food to how we build our cities and manage our waste. It's about more than just reducing harm; it's about actively enhancing the health of our planet.

The climate positive approach is inherently regenerative. It's rooted in the understanding that we need to work with nature, not against it. This means leveraging natural processes to capture and store carbon, such as reforesting degraded land and restoring peatlands and coastal ecosystems. It also means harnessing the power of technology to engineer solutions, such as direct air capture machines that can suck carbon dioxide straight out of the atmosphere.

Being climate positive also implies transforming our economies and societies. It requires shifting from a linear 'take-make-dispose' model to a circular one, where waste is minimized and resources are continually reused and recycled. It demands that we consider not only the immediate outputs of our actions but also their long-term impacts.

On the surface, the goal of climate positivity may seem daunting. After all, our planet's climate has already been significantly altered, and many of the processes driving this change are deeply entrenched in our societies and economies. But we must also recognize the potential this concept holds. By striving for climate positivity, we can help create a world that is not just sustainable, but thriving.

Climate positivity represents a fundamental paradigm shift. It reframes climate change not just as a crisis to be mitigated, but also as an opportunity for positive transformation. It calls for a future where we don't merely survive, but actively enhance the environment that sustains us.

Climate positivity is a bold, ambitious concept, and translating it into reality will require concerted effort across all sectors of society. It will require innovation and creativity, as well as the courage to challenge established norms and rethink our relationship with the planet. Yet, as daunting as this task may seem, we must remember that the stakes could not be higher. The choices we make today will shape the world of tomorrow, and in the face of climate change, striving for anything less than a positive future is simply not an option.

Case Studies: Success Stories of Climate Positivity

In recent years, we have seen a significant increase in the number of individuals, communities, companies, and even entire countries striving to go beyond sustainability and carbon neutrality to achieve climate positivity. It is a burgeoning movement fueled by the realization that we need to do more than just slow down the pace of environmental damage; we must actively work to reverse it. To illuminate this further, let's delve into a few success stories that exemplify the climate positive approach.

Our first case study takes us to Sweden and a small company called "Plantagon." Plantagon has combined urban farming with energy efficiency in a novel way that showcases a climate positive approach. They design and operate vertical urban greenhouses that not only produce locally sourced food, but also recycle excess heat and CO_2 from the city buildings in which they are housed. The result is a circular system that reduces both food and carbon miles, while also actively contributing to urban cooling and air quality. Plantagon serves as a model of how the integration of urban planning and agriculture can create climate positive solutions in our cities.

Next, we turn our attention to Costa Rica, a country that is leading the way in demonstrating how national policy can drive climate positivity. Costa Rica has committed to becoming a carbon-neutral country by 2021 and a decarbonized country by 2050. Its focus is not just on reducing emissions but also on enhancing carbon sinks through reforestation and agroforestry. Costa Rica's Payment for Environmental Services program, for example, rewards farmers and landowners for preserving forests and planting trees, thereby contributing to the removal of additional CO_2 from the atmosphere.

Meanwhile, on the other side of the world, the Maori tribe of Ngāi Tūhoe in New Zealand is redefining what a climate positive building can be. The Te Kura Whare, the tribe's new Living Building, is a testament to indigenous knowledge and sustainable technology. Constructed with local, non-toxic materials and powered by solar energy, the building produces more energy than it consumes. It also incorporates a rainwater collection system and a composting toilet system, demonstrating how the principles of a circular economy can be applied at the architectural level.

Lastly, let's look at Google, a tech giant leading the climate positive movement in the corporate sector. In 2020, Google achieved a significant milestone by eliminating its entire carbon legacy, effectively becoming carbon neutral for all of its operational history. But the company didn't stop there. It is now striving to operate entirely on carbon-free energy by 2030, across all its data centers and campuses worldwide. Google's ambition extends to driving the wider availability of carbon-free energy on electrical grids, showing how corporations can use their influence to drive systemic change.

Each of these cases exemplifies a different facet of climate positivity. From urban farms and sustainable buildings to national policies and corporate commitments, they show that the concept of climate positivity can be applied at multiple levels and in myriad ways. These stories are proof that the transition to a climate positive world is not only possible, but it's already happening.

The path to climate positivity is not without its challenges. It demands innovative thinking, determined action, and a willingness to challenge the status quo. But as these success stories show, the rewards are manifold. A climate positive approach offers a way to not only mitigate the impacts of climate change but also to create a more resilient, equitable, and sustainable world. It is a path that leads us towards a future where humans and the planet can thrive together, a future that we can all aspire to.

The Role of Technology in Driving Climate Positivity

As we explore the role of technology in driving climate positivity, we must first understand what it truly means to be climate positive. The term "climate positive" refers to an action or behavior that goes beyond achieving net-zero carbon emissions to create an environmental benefit by removing additional carbon dioxide from the atmosphere.

Today, it's becoming increasingly clear that addressing the challenges of climate change requires more than simply reducing our carbon footprint. It also involves utilizing technology to reverse the adverse effects of centuries of industrialization, deforestation, and pollution. We need to leverage our technological advancements not just to sustain, but to restore our planet.

Now, how can technology play a role in driving climate positivity?

One of the most prominent areas where technology is making a difference is in renewable energy. Solar and wind technologies have advanced significantly over the past few decades, to the point where they are now competitive with, and in many cases cheaper than, fossil fuels. As the cost of renewable energy continues to drop, it becomes an increasingly viable solution for businesses and individuals. This shift towards renewable energy sources not only reduces our reliance on fossil fuels but also helps to remove greenhouse gases from the atmosphere.

Advancements in battery technology have also been crucial. Improved energy storage capacity is allowing for the more effective use of renewable sources, by storing energy when production is high (such as sunny or windy days) and distributing it when production is low. The development of more efficient, longer-lasting, and eco-friendly batteries is a key component in the transition towards a more sustainable energy grid.

Digital technology also plays a significant role. From smart grids that optimize energy use to machine learning algorithms that predict and adapt to weather patterns for more efficient farming practices. Not to mention the use of AI and big data in climate modeling and forecasting, which provides us with crucial information for planning and adaptation.

Another promising field is carbon capture and storage (CCS) technology. This involves capturing carbon dioxide at its emission source, such as power plants, and storing it underground, preventing it from being released into the atmosphere. While still a developing technology, CCS has the potential to be a significant tool in our climate positive arsenal.

There's also the emerging field of carbon removal technologies, or "negative emissions technologies." These include direct air capture, bioenergy with carbon capture and storage (BECCS), and enhanced weathering, among others. Each offers a different method for actively removing CO_2 from the atmosphere.

Moreover, we have biotechnology providing us with ways to create biofuels, engineer crops to withstand changing climate conditions, and even utilize organisms to capture and store carbon.

The possibilities are indeed vast. Each technology presents us with an opportunity to shift our practices, industries, and economies in a more sustainable and restorative direction. However, it's important to remember that technology alone is not a silver bullet. Each of these tools comes with its own set of challenges, and the path to wide-scale implementation can be fraught with hurdles, from economic factors to social acceptance and policy support.

Nonetheless, the potential of technology to drive climate positivity is undeniable. The innovation, creativity, and problem-solving capacities that have propelled technological advancement are the same qualities we need to address the climate crisis. It is through harnessing these technologies and combining them with necessary systemic changes that we can hope to not only mitigate the impacts of climate change but also create a climate positive future.

As we continue our journey through the age of adaptation, it becomes clear that technology will play an increasingly vital role in shaping our world and our minds towards climate positivity. The challenge ahead is formidable, but armed with knowledge, determination, and technological innovation, we can face it head-on. The narrative of climate change is still being written, and with the help of technology, we have the power to steer it towards a positive ending.

Chapter 7: The Age of Adaptation: Human Resilience Amidst Climate Change

The Psychology of Adaptation: How Minds are Evolving

In this chapter, we dive into the deepest corners of the human mind, exploring the ways in which it is being reshaped in response to the realities of climate change. Adaptation, at its core, is a psychological process. The human mind, with its incredible plasticity and adaptive capabilities, is constantly evolving to face new challenges. And climate change, undeniably, presents one of the greatest challenges our species has ever encountered.

We begin by understanding the concept of psychological adaptation. In a broad sense, this refers to the cognitive and emotional adjustments people make in response to changes in their environment. It can involve changes in perceptions, attitudes, beliefs, and behaviors, and is a crucial component in our ability to deal with stress, adversity, and uncertainty.

Now, how does this process of psychological adaptation manifest in the context of climate change?

Firstly, there's the shift in perceptions. Climate change, once a distant, abstract concept for many, has become an immediate, tangible reality. Rising temperatures, more frequent and severe extreme weather events, loss of biodiversity, and shifting landscapes are making the effects of climate change increasingly visible. This heightened awareness is leading to a shift in how we perceive our relationship with the environment, prompting a more urgent sense of responsibility and stewardship.

Attitudes towards climate change are also evolving. There's a growing recognition of the need for urgent action, not only among scientists and policymakers but also among the public. This shift in attitudes is driving changes in behavior, with people increasingly making more sustainable lifestyle choices, from dietary changes and reduced consumption to activism and advocacy.

Beliefs are another important facet of psychological adaptation. As the realities of climate change become more apparent, many of our previously held beliefs are being challenged. For instance, the belief in infinite growth and consumption, a cornerstone of our current economic system, is becoming increasingly untenable. Beliefs about justice, equality, and intergenerational responsibility are also coming to the forefront, reshaping our moral and ethical landscapes.

Alongside these shifts, the climate crisis is also necessitating new coping strategies. We are having to navigate feelings of fear, grief, and anxiety, a phenomenon referred to as "eco-anxiety." Learning to manage these emotions is a crucial part of psychological adaptation, allowing us to stay engaged and proactive in the face of overwhelming challenges.

Yet, it's not all doom and gloom. The process of psychological adaptation also involves fostering resilience and hope. Embracing the concept of "active hope," where hope is seen not as a passive wish for a better future but as a capacity to envision and work towards desired outcomes, is vital. This shift helps to galvanize action and maintain momentum in the face of daunting odds.

Another positive aspect of this psychological shift is the emergence of a sense of global community and solidarity. Climate change is a global problem that requires collective action. The shared experience of facing this crisis is fostering a sense of connection and unity among people across the globe, breaking down barriers of nationality, race, and culture.

Our minds are evolving in response to climate change. This psychological adaptation is multifaceted, involving shifts in perceptions, attitudes, beliefs, and coping strategies. It's a complex, ongoing process that's fraught with challenges, yet also ripe with opportunities for growth, connection, and resilience.

As we navigate through the Age of Adaptation, it's crucial to recognize and nurture this psychological evolution. After all, our minds are the greatest tool we have in facing the climate crisis. By understanding and fostering this process of psychological adaptation, we can enhance our resilience and capacity to not just survive, but thrive in a changing world.

Social Adaptations: From Sustainable Communities to Climate Education

As we dive deeper into the era of climate change, the concept of adaptation goes beyond just individuals—it seeps into the fabric of our societies. Communities, cities, and entire societies are pivoting, learning, and innovating to confront the reality of a warming world. In this part of the chapter, we explore the manifold ways in which our societies are adapting to climate change, from the emergence of sustainable communities to the reshaping of education.

Let's begin with the concept of sustainable communities. Increasingly, we're seeing people come together to create local systems that are resilient and regenerative—capable of withstanding climate-induced stressors and also contributing to climate mitigation. This involves a rethinking of virtually every aspect of community life.

For instance, sustainable communities are turning to local, organic agriculture, reducing reliance on food imports and associated carbon emissions, while promoting biodiversity and soil health. Energy systems are being reimagined, with communities embracing renewable sources such as solar and wind. In the built environment, sustainable design principles are being applied to create buildings that are energy-efficient, resilient, and in harmony with the local climate and ecology.

But it's not just about tangible infrastructure—sustainable communities are also about social innovation. They are spaces where new forms of cooperation, governance, and social interaction are being explored. By fostering a sense of connectedness and shared purpose, these communities are demonstrating that the transition to sustainability is not just a technical challenge, but a social one.

Turning to education, climate change is reshaping curricula around the world. The goal is not just to inform students about the science of climate change, but to equip them with the skills and attitudes needed to navigate and influence a world undergoing rapid environmental change. For example, climate education is moving towards a more interdisciplinary approach, recognizing that understanding and addressing climate change requires knowledge in areas ranging from ecology and earth sciences to economics, politics, and ethics. Emphasis is being placed on systems thinking, to help students grasp the complexity and interconnections of the global climate system.

Furthermore, climate education is also focusing on fostering a sense of agency and empowerment. Rather than inducing fear and despair, the goal is to inspire students to become active participants in creating a more sustainable future. This includes not just formal education but also informal and community-based learning experiences that connect climate issues to the realities of students' lives.

In sum, social adaptations to climate change are wide-ranging and dynamic. Sustainable communities and climate education are two prominent examples of how societies are transforming in response to the challenges and opportunities of climate change. As we navigate this era of profound change, these social adaptations offer valuable lessons and insights, demonstrating that with creativity, cooperation, and courage, we can build societies that are not just resilient, but regenerative—able to thrive in a changing climate while also contributing to a healthier, more equitable world.

Policy Adaptations: The Evolution of Climate-Smart Politics

Climate change is a defining issue of our time, and it is reshaping not just our physical world, but also our political landscapes. In response to the escalating climate crisis, we're witnessing the evolution of "climate-smart politics," which refers to political strategies, policies, and leadership that recognize, incorporate, and prioritize the realities of climate change. This part of the chapter will delve into how policy adaptations are shaping the trajectory of climate-smart politics.

Firstly, the acknowledgement of climate change as a cross-cutting issue is becoming central to policy-making. Previously compartmentalized as a purely environmental issue, climate change is now recognized as a pervasive force that influences every policy area— from public health to national security, from economic planning to social justice. This has led to the emergence of integrated, cross-sectoral policy frameworks that strive to address climate change in conjunction with other societal goals. Countries are working towards embedding climate considerations within all policy spheres, shaping an overarching strategic approach that aligns with climate change mitigation and adaptation.

Secondly, the focus of policy adaptations has expanded from mitigation to include adaptation. While cutting emissions to prevent further climate change is absolutely crucial, the impacts of already-changed climate are here, and they are growing. Recognition of this fact has led to an upswing in policies that aim to boost societal resilience and facilitate adaptation to the new climate realities. These range from infrastructure regulations that account for rising sea levels and extreme weather, to social policies that support communities, particularly vulnerable ones, in becoming more climate-resilient.

Another major evolution is the emphasis on climate justice in policy adaptations. As the climate crisis unfolds, it's becoming increasingly clear that it's the underprivileged who are often hit the hardest. In response, there's a growing movement to ensure that climate policies are not just effective, but also equitable. This means focusing on those most vulnerable to climate impacts and addressing the systemic inequalities that exacerbate climate vulnerabilities. It's about making sure that the transition to a low-carbon, climate-resilient society is a just transition, one that doesn't leave anyone behind.

International cooperation is another critical dimension of climate-smart politics. Climate change is a global problem that respects no borders, and its solution necessitates global action. We're seeing this reflected in the evolution of international climate agreements, from the Kyoto Protocol to the Paris Agreement, each bringing more countries into the fold and striving to increase ambition and accountability. At the same time, international climate finance is being mobilized to support developing countries in pursuing low-carbon, climate-resilient development.

Lastly, climate-smart politics involves actively engaging with a diverse array of stakeholders. Governments are increasingly recognizing that to effectively address climate change, they need to involve civil society, businesses, academia, and the public in the policy process. Whether it's through public consultations on climate plans, partnerships with private sector for green innovation, or collaboration with researchers for climate risk assessment, multi-stakeholder engagement is becoming a cornerstone of climate-smart politics.

To sum up, policy adaptations are driving the evolution of climate-smart politics, marked by integration, resilience, justice, cooperation, and inclusiveness. As the climate crisis deepens, the need for such politics becomes ever more urgent.

With their ability to shape societal trajectories, these policy adaptations and the climate-smart politics they engender will play a pivotal role in our collective ability to navigate the tumultuous era of climate change.

Chapter 8: Niche Perspectives on Climate Change and Adaptation

Climate Change and Intersectionality: A Unique Lens

In this section, we shall explore a unique lens through which we can view the impacts of climate change and how we can adapt to them – intersectionality. For those unfamiliar with the term, intersectionality refers to the interconnected nature of social categorizations such as race, class, and gender, creating overlapping systems of discrimination or disadvantage. The intersectionality framework can provide us with a rich and nuanced understanding of the differential impacts of climate change and the need for inclusive and just adaptation strategies.

The intersectional perspective acknowledges that climate change does not impact everyone in the same way. The effects are felt differently, depending on where one stands at the crossroads of various social, economic, and political identities. For instance, it's well-documented that people living in poverty are disproportionately affected by climate change. But an intersectional view would also consider how, within that group, women are often more vulnerable due to gender inequalities that limit their resources and adaptive capacities.

This differential impact is due to a host of intersecting factors that can magnify a person's or a group's vulnerability to climate change. Factors such as race, socio-economic status, gender, age, and geographical location can all intersect to create layered experiences of climate change.

For instance, consider a low-income, elderly woman living in a flood-prone coastal community. Her experience of climate change is shaped not just by one, but by multiple intersecting identities – her socio-economic status makes her less able to afford protective measures; her age may limit her physical ability to cope with extreme weather events; her gender may make her more susceptible to violence or neglect during disasters; and her location exposes her to rising sea levels and intensified storms.

But intersectionality isn't just about vulnerability – it's also about resilience. It highlights the ways in which different groups, facing multiple forms of discrimination, have developed unique coping strategies that can be harnessed for climate change adaptation. Indigenous communities, for example, have a wealth of traditional ecological knowledge that has allowed them to adapt to changing environmental conditions for centuries. Women in many societies are the primary caregivers and food providers, roles that give them unique insights and skills in managing resources and dealing with crises.

So, how can we apply an intersectional lens to climate change adaptation?

Firstly, intersectionality can help ensure that climate change policies and actions are inclusive and equitable. It can guide us to look beyond averages and consider the range of different experiences and needs. It challenges us to ask: Who is most vulnerable? Who is being left out? How can we ensure that our actions do not exacerbate existing inequalities but contribute to social justice?

Secondly, intersectionality can inform more effective and sustainable adaptation solutions. By acknowledging the different ways in which people experience and respond to climate change, we can design strategies that are more in tune with the realities on the ground. It allows us to recognize and value the diverse knowledge and skills that different groups bring to the table, and to build on these strengths in our adaptation efforts.

Climate change, as a global issue, has far-reaching impacts, and these impacts are not always apparent when viewed from a singular perspective. While a macro view of climate change can give us a broad understanding of its impacts, an intersectional perspective provides a more detailed and nuanced understanding of these impacts at a micro level. This detail is critical for effective planning and decision making.

Applying intersectionality in climate change research involves a departure from a 'one-size-fits-all' approach to understanding and responding to climate change. It involves taking into account the unique ways that various social, political, and economic systems interact with environmental changes. This, in turn, provides a more precise and detailed understanding of both the impacts of climate change and the most effective strategies for mitigating these impacts.

Let's consider an example of how intersectionality might inform policy. In many societies, women are primarily responsible for collecting water. As climate change disrupts water supplies, this task becomes increasingly difficult and time-consuming. A policy response that simply improves overall access to water might overlook the gendered nature of this impact. By contrast, an intersectional approach would recognize this issue and could lead to a more targeted solution, such as efforts to reduce the time spent on water collection or to engage women in water management decision-making.

Similarly, intersectionality can help us understand the impact of climate change on marginalized racial and ethnic communities. These groups often have fewer resources to cope with climate change and may face social and political barriers to implementing adaptive strategies. An intersectional perspective can reveal these unique challenges and lead to policies that address systemic barriers to adaptation.

Moreover, understanding intersectionality can also enhance our efforts in climate change communication.

Recognizing the diverse experiences and perceptions of climate change among different groups can help tailor communication strategies that resonate with a wide range of audiences. For example, messaging that connects climate change to issues of social justice may be more effective among audiences concerned with these topics.

In essence, an intersectional perspective provides a tool for understanding the complex ways that different groups of people experience climate change. This perspective can improve the effectiveness of climate change research, policy, communication, and action by illuminating the diverse impacts of climate change and the diverse strategies for addressing these impacts.

By incorporating intersectionality into our understanding of climate change, we can move closer to a world in which climate change policies and interventions are both effective and equitable. By recognizing and embracing our diversity, we can find strength and resilience in the face of a changing climate. The task at hand is complex, but so is the world we inhabit – and it is this complexity that, if understood and harnessed, can guide us towards a resilient and sustainable future.

Climate Change in Literature and Art: The Power of Creative Expression

Let's explore how literature and art have become powerful mediums for expressing the impacts and experiences of climate change.

Climate change is not only a scientific issue but a cultural one as well. Literature and art have a unique ability to convey the emotional and experiential dimensions of climate change, providing a deeper understanding of its impacts and complexities.

The use of literature to explore climate change has given birth to a new genre known as "cli-fi" or climate fiction. This genre employs narrative and characterization to illustrate the possible futures and ethical dilemmas we face as the climate changes. Cli-fi allows us to imagine how societies might adapt, or fail to adapt, to a changing environment, opening our eyes to the human stories behind the facts and figures.

A striking example of cli-fi is Paolo Bacigalupi's "The Water Will Come." The novel, set in a future where sea levels have risen significantly, showcases the implications of our current actions on future generations and emphasizes the reality of our current trajectory if we fail to mitigate climate change. By transporting us into a possible future, cli-fi novels like this one evoke a sense of urgency and offer a dramatic exploration of the moral and ethical issues at stake.

On the other hand, poetry has its unique way of addressing climate change. Through its emotive and condensed language, poetry has the capacity to encapsulate the profound impacts of climate change. For example, "Rising," by Elizabeth Rush, is a collection of poems that bears witness to the communities affected by sea-level rise, illuminating the deep and personal connections people have with their threatened environments.

Turning to the visual arts, artists worldwide have utilized their platforms to bring the reality of climate change to life. From striking installations that visualize rising sea levels, to photography that captures the devastating impacts of extreme weather events, visual art engages the viewer's senses and emotions, making the abstract concept of climate change palpable.

Artists like Olafur Eliasson and Minik Rosing have made tangible the abstractness of climate change data. In their installation "Ice Watch," giant blocks of ice from Greenland were placed in public spaces, allowing people to watch them melt - a visual and visceral demonstration of the effects of global warming.

The power of art lies in its ability to evoke emotion, elicit empathy, and create a sense of urgency. This makes it an incredibly potent tool for raising awareness and promoting action on climate change.

Moreover, creative expression through literature and art provides a means for diverse voices to be heard. Through their works, writers and artists from marginalized communities can highlight their unique experiences of climate change, fostering a broader and more inclusive understanding of its impacts.

Literature and art have a profound role in shaping our cultural understanding of climate change. Through their capacity to convey complex emotions and experiences, they bring the human dimension of climate change to the fore, fostering empathy, understanding, and action. As we navigate the complexities of the Anthropocene, these creative mediums continue to serve as essential tools for exploring, communicating, and responding to the existential challenge of our time.

Art and literature do not merely echo the scientific facts about climate change; they breathe life into these facts, humanizing and personalizing them. They do this by engaging our senses, stirring our emotions, and stirring our collective conscience.

In the world of film, climate change has been an increasingly prevalent theme. Take, for instance, the documentary "An Inconvenient Truth" by former U.S. Vice President Al Gore. This film arguably brought climate change to the mainstream, making the science accessible and relatable to a global audience. It utilized the power of visual storytelling to reveal the urgent issue of global warming, winning two Academy Awards and inspiring a generation of climate activists.

Similarly, the science fiction film "Interstellar" uses the concept of a future Earth rendered uninhabitable by climate change as its central plot device. While it's a work of fiction, it paints a grim potential future that could result from our current environmental negligence.

In terms of visual arts, street art has also been a potent medium for climate change messaging. The internationally recognized British street artist, Banksy, for example, has often used climate change as a theme in his provocative works. One of his murals features a depiction of a girl being swept away by rising waters, a potent visual metaphor for the urgent threat of sea-level rise.

Art installations can also engage the public on a large scale. The "Climate Clock" installation in Union Square, New York, for instance, has become an iconic representation of the urgency of addressing climate change. This large-scale digital clock counts down the critical time left to prevent irreversible effects of global warming, striking a chord with the urgency of our situation.

In the realm of literature, non-fiction works such as Naomi Klein's "This Changes Everything: Capitalism vs. The Climate" and Elizabeth Kolbert's "The Sixth Extinction" have played significant roles in shedding light on the political, economic, and ecological aspects of climate change. These books, among others, have facilitated the broader comprehension of climate change's complexities.

Art and literature serve as crucial outlets for the expression of fear, frustration, and hope in the face of climate change. They allow us to grapple with the emotional weight of climate change, something raw data alone cannot accomplish. More importantly, they have the power to catalyze conversations, shape public opinion, and influence policy-making, making them indispensable in the fight against climate change.

In essence, literature and art are far more than forms of entertainment in the context of climate change. They serve as both mirrors and windows—mirrors that reflect our reality and windows that offer us a glimpse into the potential futures we could face. By making the invisible visible and the distant feel closer, they make us confront the uncomfortable truths about climate change, ultimately pushing us towards action and adaptation.

The Role of Indigenous Knowledge in Climate Adaptation: An Undervalued Resource

In this journey towards understanding climate change and its implications, it is crucial that we discuss the significance of indigenous knowledge systems. These systems, which represent thousands of years of human interaction with the natural environment, provide us with an indispensable resource for understanding, mitigating, and adapting to the impacts of climate change.

Indigenous communities worldwide have long observed and responded to subtle shifts in their environments, developing highly sophisticated ways of understanding their ecosystems. Such intimate, multi-generational knowledge can provide invaluable insights into sustainable land and water management practices that have sustained diverse ecosystems over millennia.

Indigenous knowledge systems offer holistic worldviews where humans are a part of, rather than apart from, the natural world. This perspective contrasts with the dominant anthropocentric view in many societies today, which prioritizes human needs and wants above the health of our environment.

In the Arctic regions, for example, indigenous peoples such as the Inuit have a profound understanding of ice and wildlife behaviors, accumulated over generations. This knowledge, including how to predict weather and identify safe ice for travel, plays a critical role in their survival. With the onset of climate change, their adaptive strategies provide critical lessons for resilience in an increasingly unpredictable world.

Indigenous agricultural practices, too, hold keys to climate resilience. For instance, the Indigenous people in Peru's

Andean region have, for centuries, cultivated a diverse range of potato varieties suited to different microclimates. This biodiversity serves as a buffer against climate variability, ensuring food security amidst changing conditions. In contrast, modern agriculture's focus on monocultures is highly susceptible to climate change and pests, underscoring the need to integrate indigenous wisdom into contemporary farming practices.

Moreover, indigenous communities often have a deep cultural understanding of climate change, viewing it as a symptom of the profound disconnection between humans and nature. For instance, many Pacific Island cultures have a strong "ocean identity," recognizing the intricate linkages between their lives, the sea, and the broader ecosystem. As they face the existential threat of rising sea levels, their voices are critical in global conversations about climate justice and adaptation.

Yet, despite the value of indigenous knowledge, it is often overlooked or outright dismissed in climate science and policy. This dismissal can be traced to power imbalances, institutional biases, and the marginalization of indigenous peoples globally. As a result, we lose out on essential wisdom that could inform more sustainable and adaptive practices.

Indigenous knowledge is not a panacea for the climate crisis, nor should it be romanticized or appropriated without respecting indigenous rights and sovereignty. However, it offers vital insights into living more harmoniously with our environment. By recognizing the value of indigenous knowledge, we can foster a more inclusive, effective, and culturally diverse approach to climate adaptation.

Promoting dialogue and collaboration between scientists, policymakers, and indigenous communities is critical. Collaborative efforts like the Intergovernmental Science-Policy Platform on Biodiversity and Ecosystem Services (IPBES) serve as promising models, integrating indigenous knowledge into global assessments on biodiversity and ecosystem services.

In the fight against climate change, we need all hands on deck. As we move forward, it's crucial that we appreciate the wisdom of indigenous peoples and ensure their perspectives are integrated into climate adaptation strategies. Only by doing so can we truly hope to navigate the challenging path towards sustainability and resilience in our changing world.

Chapter 9: Preparing for the Future: The Path Ahead

Individual Actions: What You Can Do

Our journey through the climate crisis, exploring its implications from varied vantage points, brings us to a crucial juncture: individual action. Climate change may be a global issue, but its solution lies in the hands of every individual on this planet. The choices we make, the actions we take, the voices we raise—all play a crucial role in our collective response to climate change.

As an individual, understanding your carbon footprint is a crucial first step towards climate action. Your carbon footprint is the total amount of greenhouse gases, primarily carbon dioxide, emitted due to your actions and lifestyle. It includes the emissions from the energy you use at home, the transport you take, the food you eat, and the products you buy. Many online tools can help you calculate your carbon footprint, giving you a clear picture of your impact on the environment.

Once you understand your carbon footprint, you can identify areas where you can reduce it. Energy efficiency, for instance, is an effective way to cut down emissions. Simple changes like switching to LED bulbs, insulating your home, or installing solar panels can significantly reduce your energy use and carbon emissions.

Similarly, rethinking your transport options can make a difference. Can you walk, cycle, or take public transport instead of driving? If you need a car, could it be an electric or hybrid one? For long distances, is it possible to train instead of flying?

Dietary choices also play a big role. Meat and dairy products, particularly from industrial farming systems, have a much higher carbon footprint than plant-based foods. You don't have to go vegan overnight, but even reducing your meat and dairy intake and choosing locally sourced foods can make a significant difference.

It's not just about reduction, though. As individuals, we can take proactive steps to support renewable energy, reforestation, and other climate-positive initiatives. Plant trees, install a rainwater harvesting system, start composting—there are countless ways you can contribute to a more sustainable world.

Crucially, it's not just about what you do—it's also about the message you send. By adopting sustainable practices, you show others that a greener lifestyle is possible and desirable. You influence your friends, family, and community, and contribute to a broader cultural shift towards sustainability.

But individual actions are not enough in isolation. It's essential to engage with politics, to use your voice and your vote to push for climate policies. Contact your representatives, vote for leaders who take climate change seriously, participate in peaceful protests, support organizations that fight for climate justice—these are all ways to make a difference.

Lastly, educate yourself and others. Knowledge is power, and in the case of climate change, knowledge can mean survival. Learn about climate science, keep up-to-date with the latest research, and spread that knowledge. Education breeds understanding, and understanding breeds action.

Taking individual action on climate change may seem daunting, but it's both necessary and empowering. When each of us takes responsibility for our actions and their impact on the planet, we're not just preparing for the future—we're actively shaping it. The path ahead may be fraught with challenges, but with every step we take, we bring a sustainable future within reach.

There is also an essential need to understand and participate in the circular economy – an economic model that aims to eliminate waste and encourage the continual use of resources. This model can be adopted on individual levels in our everyday choices. Embrace the concept of 'reduce, reuse, recycle' in your daily life. This could be as simple as opting for reusable shopping bags over single-use plastic, mending or upcycling clothes instead of discarding them, composting organic waste, or choosing products with less packaging.

Furthermore, investing in green businesses and initiatives can be another significant step. As consumers, we have the power to support companies that prioritize sustainability, whether it's through the products they create or their operational practices. By making informed choices and supporting these companies, we can indirectly contribute to the reduction of carbon emissions and the growth of the green economy.

On the other hand, we can also hold accountable the corporations that are not environmentally conscious. Advocacy is a powerful tool, and public pressure can often lead to substantial changes in corporate behavior. Using your platforms, no matter how big or small, to voice concerns and demand accountability is another impactful form of individual action.

Financial investments also hold power. Consider green investments – investing in funds that prioritize environmental, social, and governance (ESG) criteria. It's a way to ensure your money supports businesses that are doing their part to combat climate change. Not only can this provide you with financial returns, but it also helps grow businesses that are contributing positively to the environment.

In the realm of education, encouraging climate literacy in schools is crucial. Supporting initiatives that incorporate climate education into the curriculum will ensure the next generation is better equipped to deal with the challenges ahead. Knowledge passed down to the younger generation can induce a sense of responsibility and urgency to act against climate change.

Remember that it's okay to start small. The enormity of the climate crisis can be overwhelming and it's easy to feel like the actions of one person can't make a difference. But the reality is, they do. Each action we take adds up, and together, we can make a difference.

Individual actions are the building blocks of societal change. Climate change might be a global problem, but it has local solutions that start with you and me. The path ahead is challenging, but not insurmountable. By taking individual actions, we are not only helping to mitigate the effects of climate change, we are also building a resilient society that will be prepared to adapt to the changes that lie ahead. It's an ongoing process, but one that holds the promise of a sustainable future. Remember, our actions today will determine the climate of tomorrow.

Policy Recommendations: Towards a Climate-Resilient Future

In the quest to navigate our warming world and to shape a climate-resilient future, policy recommendations play a crucial role. These guidelines help nations to create effective, integrated, and holistic strategies that not only mitigate climate change impacts but also promote adaptation, ensuring communities are resilient in the face of climate uncertainty. Here, we present several vital policy recommendations for a future ready to face the challenge of climate change.

To start, we need policies that advance green technology and renewable energy. As one of the most impactful sectors contributing to greenhouse gas emissions, the energy sector's transformation is paramount. Policies should encourage research and development into renewable energy sources such as solar, wind, hydro, and geothermal. Subsidies for green technologies can be a significant catalyst in this sector, making renewable energy competitive with traditional fossil fuels. Policymakers must also facilitate grid infrastructure development to handle renewable energy and storage technologies.

Next, energy efficiency policies are essential. It's not just about how we generate our energy, but also how efficiently we use it. Energy conservation efforts, like the implementation of energy-efficient appliances, insulation, and lighting in homes and businesses, can significantly reduce overall energy demand. Regulatory measures, such as mandating energy performance labels on appliances and setting minimum efficiency standards, are practical strategies to accomplish this.

Moreover, transitioning to a circular economy should be high on the policy agenda. This model, which emphasizes reducing, recycling, and reusing, promotes economic growth while minimizing waste and resource use. Policies can incentivize businesses to design products that are long-lasting, repairable, and recyclable. Waste management regulations can be updated to prioritize recycling and composting, and discourage landfill use.

In agriculture, policies should be geared toward promoting sustainable farming practices. The encouragement of organic farming, permaculture, agroforestry, and practices that enhance soil health can not only reduce carbon emissions but also make our food systems more resilient to climate impacts. Policies could take the form of subsidies for sustainable practices, training programs for farmers, and research into new and innovative farming methods.

Public transportation and urban planning policies also hold significant potential. Urban environments are often significant contributors to greenhouse gas emissions, but they also present opportunities for energy savings and emission reductions. Implementing policies that encourage the use of public transport, walking, and cycling can reduce reliance on private vehicles. Additionally, urban planning policies should promote green spaces, which can absorb CO_2, reduce urban heat island effects, and improve residents' overall wellbeing.

In the realm of conservation, policies should be enacted to protect and restore our ecosystems, which serve as natural carbon sinks. Forests, wetlands, and mangroves sequester significant amounts of carbon and provide invaluable services such as flood control and habitat provision. Implementing strict anti-deforestation policies, creating protected areas, and funding restoration projects can preserve these critical ecosystems.

Moreover, financial regulations need to account for climate change. This means encouraging green investments and discouraging investments in fossil fuels. Policies can mandate that financial institutions assess and disclose the climate risk of their portfolios. Tax incentives can be given to green investments, and carbon pricing can make it costly for businesses to pollute.

Lastly, it's vital to integrate climate education into our school curricula. Understanding the causes, impacts, and solutions to climate change is critical for fostering future generations who can innovate and implement climate solutions. Policymakers should thus promote the inclusion of climate science in all levels of education.

These policy recommendations, while not exhaustive, provide a starting point towards a climate-resilient future. They are guidelines that help pave the path ahead. Nevertheless, it's essential to remember that the implementation of these policies will vary based on regional and national contexts, and thus, flexibility and adaptation are key. Each policy, every law, and regulation that steers us toward a greener, more sustainable future is like a stone paving the way of our path towards climate resilience. It's not a one-size-fits-all solution, but a mosaic of different approaches suited to local conditions and needs.

In this context, it's crucial to involve local communities in the policymaking process. Policies should be culturally sensitive, socially inclusive, and beneficial to local economies. This way, the path to climate resilience becomes a shared journey, backed by the collective will and participation of the people they affect.

Policies for climate resilience should also take into account the needs and vulnerabilities of marginalized groups, such as indigenous communities, women, and the poor. These populations often suffer disproportionately from the impacts of climate change, yet they also hold unique insights, knowledge, and skills that can greatly contribute to climate resilience. An equity-focused approach to policy-making can ensure that all voices are heard, and that everyone has an opportunity to participate in shaping a climate-resilient future.

Moreover, to maximize the effectiveness of these policy interventions, international cooperation is essential. Climate change is a global problem that knows no borders, and its solutions require global action. Countries should work together to share knowledge, technology, and resources. International agreements, such as the Paris Agreement, can set shared goals and frameworks for climate action. However, these should be complemented with robust mechanisms for accountability and support for countries that need it the most.

It's also essential to integrate climate change considerations into all policy areas, a concept known as 'climate mainstreaming.' For instance, policies for economic development, public health, or infrastructure should all consider the impacts and risks associated with climate change. This way, we can ensure that our path towards climate resilience is pursued across all sectors of society, and not just those directly related to the environment.

Policymakers must not forget the importance of monitoring and evaluation. Once implemented, policies should be regularly reviewed to assess their effectiveness and make necessary adjustments. Data collection and climate modeling can help track progress and predict future scenarios. Such mechanisms ensure that our path towards climate resilience is not only proactive but also adaptive, ready to respond to new challenges and opportunities that may arise.

The path to a climate-resilient future requires a comprehensive, inclusive, and flexible policy approach. It's about embracing green technology, promoting sustainable practices, conserving our ecosystems, advancing climate education, and much more. It's about listening to diverse voices and integrating climate action across all policy areas. Most importantly, it's about working together, both locally and globally, to create a sustainable and resilient world. As we navigate this path, every step we take is a testament to our ingenuity, resilience, and collective will. Through informed and responsible action, we can shape a future that thrives amidst change, a world that is not just enduring, but flourishing in the Age of Adaptation.

Research Directions: The Unexplored Terrain in Climate Change and Adaptation

Research is fundamental to our understanding and response to climate change. As we have seen in the previous chapters of this book, the scientific consensus about climate change and its impacts is strong. Still, many aspects require further investigation, and new areas of research are emerging as the climate crisis evolves. This final section of Chapter 9 will delve into some of these unexplored terrains in climate change research and adaptation.

First and foremost, the complexity and interconnectedness of the Earth's climate system mean that there are many elements we still don't fully understand. For instance, how feedback mechanisms—processes that can either amplify or diminish the effects of climate change—work in detail remains a topic for further exploration. Feedback mechanisms such as the melting of polar ice, which reduces the Earth's reflectivity and leads to further warming, can have significant impacts on our climate. Understanding these processes better can refine our climate models and improve our predictions for the future.

Another area that requires further exploration is the nexus of climate change, biodiversity, and ecosystems. While we know that climate change has significant effects on biodiversity, we're just beginning to unravel the complex relationships between species and their changing environments. Similarly, the role of ecosystems as carbon sinks and their potential to mitigate climate change is an exciting and growing field of research. Further work in this area can help us identify ways to conserve biodiversity and leverage natural ecosystems in our fight against climate change.

In the realm of human systems, the psychological and behavioral aspects of climate change are ripe for exploration. As we've discussed earlier in this book, understanding how people perceive and respond to the threat of climate change is crucial for promoting sustainable behavior and supporting climate adaptation efforts.

Yet, many questions remain. For example, what motivates people to act on climate change, and how can we effectively communicate about climate change to inspire action and overcome skepticism? These and related questions present exciting opportunities for interdisciplinary research, bridging the gap between the environmental sciences, psychology, communication studies, and other fields.

Climate change adaptation strategies are another area that requires further research. While we have a growing understanding of the various strategies available to us, many questions remain about their efficacy, trade-offs, and implementation. For instance, how can we design adaptation strategies that are not just technically effective but also socially and politically feasible? How can we ensure that they are equitable, benefiting those most vulnerable to climate change? Exploring these questions can help us navigate the complexities of climate adaptation and ensure that our responses to climate change are both effective and just.

The role of technology in addressing climate change is an area of active research. Still, it's also one that has many unexplored avenues. For example, how can emerging technologies like artificial intelligence and big data be leveraged for climate action? How can we develop and scale up carbon capture and storage technologies? How can we ensure that technological solutions are accessible and beneficial to all, rather than exacerbating existing inequalities? These are crucial questions for making our fight against climate change successful and fair.

Lastly, the economic aspects of climate change present many opportunities for further research. While the economics of mitigation—such as the cost-effectiveness of different greenhouse gas reduction strategies—are well-studied, the economics of adaptation are less so. Understanding the costs and benefits of different adaptation strategies, and how these are distributed across different populations, can inform policy-making and help us prepare for a climate-changed future in an economically viable way.

While we've made significant strides in our understanding of climate change and how to respond to it, there is much terrain still to explore. These and other research directions present opportunities to deepen our knowledge, hone our strategies, and foster innovation. As we move forward in the Age of Adaptation, this spirit of exploration and curiosity will be crucial for navigating the challenges and opportunities that climate change presents.

The need for predictive climate science, which will allow us to forecast climate change impacts on local scales, is a critical area of exploration. While our current models do a good job of predicting global temperature trends, their ability to predict climate impacts at the local level—where decisions about climate adaptation are often made—is less developed. Improving these models will involve incorporating more detailed data about local landscapes, weather patterns, and human activities, among other factors. By enhancing the granularity of our predictions, we can better plan for and respond to the impacts of climate change in different regions.

Another area of ongoing research lies in the intersection of climate change and public health. Changes in temperature and precipitation patterns affect the spread of infectious diseases, air and water quality, and food security, all of which have significant implications for human health. Delving deeper into this complex interplay could offer insights that inform both public health interventions and climate adaptation efforts. For example, predictive models could help us anticipate disease outbreaks related to climate change, while a better understanding of the health benefits of climate action could bolster support for such measures.

The influence of climate change on social systems and geopolitical relations is another rich field for future exploration. Climate change is likely to exacerbate social inequalities and could lead to political conflict over resources or migration, but our understanding of these dynamics is far from complete. Research in this area can help us anticipate and mitigate the social and political fallout of climate change and design policies that promote social resilience in the face of these challenges.

Moreover, research should focus on the resilience and adaptability of our infrastructure to climate change. From power grids and transportation systems to housing and urban design, our built environment plays a crucial role in our ability to adapt to climate change. Yet, many of our infrastructures were not designed with climate change in mind. How can we retrofit existing infrastructure for a changing climate, and what principles should guide the design of new infrastructure? Answering these questions will be crucial for safeguarding our societies in the face of climate change.

Climate change also raises profound philosophical and ethical questions. Who bears responsibility for addressing climate change, and how should the burdens and benefits of climate action be distributed? How should we value non-human nature, future generations, and the non-material aspects of well-being that are threatened by climate change? These are not just abstract academic questions, but issues that have direct implications for policy-making and public discourse on climate change. Exploring these questions can enrich our responses to climate change, ensuring they are not just effective, but also ethically sound.

Further exploration is needed into the lessons we can learn from history and from different cultures in dealing with climate change. How have past societies responded to environmental changes, and what can we learn from their successes and failures? How do different cultures understand and relate to the natural world, and how can these perspectives inform our own relationship with nature? Insights from history and anthropology can offer valuable perspectives as we navigate our own age of adaptation.

The challenges posed by climate change require responses from across the sciences and humanities. As we look ahead, research in these and other areas will be crucial for informing our actions and shaping our future in a warming world. It's up to us to venture into these unexplored terrains, deepen our understanding, and translate that knowledge into tangible actions. The path ahead may be challenging, but it is also ripe with opportunities for discovery and growth. As we step into the future, let us do so with a spirit of exploration and adaptability, ready to face the complexities of climate change head-on.

Chapter 10: The Intersection of Climate Change and Public Health

Climate Change and Disease: The Rise of Vector-Borne Illnesses

Climate change has far-reaching implications, affecting not only our physical environment but also the health of populations around the world. While the adverse effects of rising global temperatures, severe weather events, and sea-level rise are evident, the correlation between climate change and public health is less apparent but equally alarming. This chapter zeroes in on one of the most insidious effects of climate change on public health - the rise of vector-borne illnesses.

Vector-borne diseases are infections transmitted by the bites of infected arthropod species, such as mosquitoes, ticks, and fleas. These diseases have been a long-standing public health problem, but climate change is exacerbating the situation by creating more favorable conditions for vectors, expanding their geographical range, and increasing the duration of transmission seasons.

One striking example is the spread of mosquito-borne diseases. Changes in temperature and rainfall patterns can affect mosquito breeding and survival rates, and in turn, alter the incidence of diseases like malaria, dengue, Zika, and West Nile virus. Warmer temperatures can accelerate the mosquito's lifecycle and the replication of the virus within the mosquito, leading to a higher transmission potential. Furthermore, increased rainfall can create more breeding sites, while extreme flooding can also increase the risk of infection by flushing out larval habitats and prompting a surge in mosquito populations.

For instance, the Aedes aegypti and Aedes albopictus mosquitoes, known transmitters of dengue, Zika, chikungunya, and yellow fever, are particularly sensitive to climatic conditions. Native to tropical and subtropical regions, these species have now been observed in parts of Europe and North America. The unprecedented 2015-2016 Zika outbreak in South and Central America, which led to a significant increase in neonatal microcephaly cases, can be partially attributed to climatic factors that favored the proliferation and geographic expansion of these mosquito species.

Ticks, another significant vector, are also expanding their range due to climate change. Lyme disease, transmitted by the black-legged tick, is the most common vector-borne disease in the United States. Warmer winters and earlier springs have allowed ticks to become active earlier in the year and expand into new regions, increasing the risk of Lyme disease.

There is also a human dimension to the problem. Socio-economic factors, urbanization, and population movement can further exacerbate the spread of vector-borne diseases. For example, urban areas with inadequate water and waste management systems can become breeding grounds for disease-carrying mosquitoes. Additionally, people displaced by extreme weather events or rising sea levels may be forced into crowded living conditions with poor sanitation, increasing the risk of disease transmission.

As we look toward the future, it becomes evident that our public health strategies need to adapt to this changing landscape. Climate-sensitive surveillance systems are crucial for early detection and timely control of vector-borne disease outbreaks. These systems should leverage weather and climate data to predict disease outbreaks and guide vector control interventions. Climate change mitigation and adaptation strategies also need to include measures to control and prevent vector-borne diseases. This could mean improving water and waste management in urban areas, promoting the use of insecticide-treated nets and vaccines where applicable, and investing in the development of new control methods.

Climate change is a complex problem that demands a multi-sectoral response. As this chapter highlights, addressing climate change is not just about reducing greenhouse gas emissions or protecting our ecosystems— it's also about safeguarding our health. The link between climate change and vector-borne diseases underscores the need for climate action to be integrated across all sectors of society, from energy and transport to agriculture and health. Only by acknowledging and addressing these interconnected risks can we hope to build a climate-resilient future.

Beyond the direct effects of vector-borne illnesses, the broader intersections of climate change and public health are manifold. Changes in air quality, shifts in food and water safety, and increased exposure to extreme weather events are additional factors in the intricate relationship between our environment and health.

In the context of air quality, climate change exacerbates air pollution, primarily by increasing ground-level ozone and particulate matter concentrations. Rising temperatures can amplify the ozone formation process, while changes in weather patterns can increase the duration and intensity of pollution episodes. The adverse health effects of air pollution include respiratory conditions, cardiovascular diseases, and premature deaths. Additionally, longer pollen seasons due to early springs and late autumns can aggravate allergies and asthma.

In terms of food and water safety, the impacts of climate change pose substantial threats to these crucial resources. Higher temperatures and changes in precipitation can contaminate freshwater sources, while extreme weather events can damage infrastructure and lead to waterborne disease outbreaks. Additionally, higher sea surface temperatures can promote the proliferation of harmful algal blooms and shellfish contamination, impacting food safety and coastal economies.

Moreover, the impacts of climate change on agricultural productivity can have far-reaching implications for food security and nutrition. Changes in temperature and precipitation patterns can affect crop yields, while an increase in extreme events can disrupt food production and supply chains. Malnutrition, especially among vulnerable populations, can lead to a multitude of health problems.

Extreme weather events—increasing in both frequency and intensity due to climate change—also bear substantial health risks. Heatwaves can lead to heat exhaustion, heatstroke, and even death, especially among the elderly, the very young, and those with pre-existing health conditions. Floods and storms can cause injuries and deaths, as well as mental health impacts such as post-traumatic stress disorder. Furthermore, these events can disrupt healthcare services when they are most needed.

Looking ahead, the health sector has a pivotal role to play in addressing the health risks of climate change. It can advocate for climate mitigation policies, given the significant health co-benefits of reducing greenhouse gas emissions. It can lead by example, reducing its carbon footprint and implementing sustainable practices. Finally, it can prepare for the future by strengthening health systems to be more climate-resilient.

Public health professionals need to be equipped with the knowledge and skills to understand and respond to these climate-sensitive health risks. This means integrating climate change and health into the medical and public health curricula. Furthermore, there is a need for more research to understand the precise mechanisms linking climate change and health outcomes, and to develop effective interventions.

Ultimately, addressing the health impacts of climate change calls for a paradigm shift: from viewing health and the environment as separate entities, to understanding them as intrinsically interconnected. A healthy planet is a prerequisite for healthy people. By protecting our planet, we are safeguarding our own health, and that of generations to come.

Mental Health in a Warming World: The Impending Crisis

As our world continues to warm, an impending crisis is emerging that requires urgent attention: the profound impact of climate change on mental health. Though less visible than melting ice caps or intensifying hurricanes, the psychological consequences of climate change are no less real and can be devastating.

To grasp the seriousness of this issue, it's vital to understand that our psychological well-being is intimately linked with our environment. Our minds are attuned to the rhythms and patterns of nature, and when these are disrupted, so too are our mental states. In other words, the health of our planet is intrinsically tied to the health of our minds.

Climate change-induced mental health impacts are multifaceted, with immediate, short-term effects and also lingering, long-term ones. These impacts range from trauma and post-traumatic stress disorder (PTSD) in the aftermath of extreme weather events, to anxiety and depression resulting from gradual environmental changes and uncertain futures. Each presents unique challenges to individuals, communities, and mental health professionals.

At the immediate level, extreme weather events such as hurricanes, wildfires, and floods—growing more frequent and severe due to climate change—can inflict psychological trauma. Survivors often experience acute stress reactions and may develop PTSD, a serious condition characterized by flashbacks, hypervigilance, and emotional numbness. Extreme weather events can also lead to displacement, a risk factor for mental health problems due to the loss of home, community, and sense of belonging.

More subtle and gradual environmental changes, such as rising temperatures and sea levels, can also undermine mental health. These slow-onset changes can lead to chronic environmental stress, manifesting as persistent anxiety, depression, and feelings of helplessness and despair, a phenomenon sometimes referred to as "eco-anxiety" or "climate grief."

Moreover, the anticipatory anxiety and existential dread about an uncertain future—a future of worsening conditions and unprecedented challenges—are notable aspects of climate change-induced psychological distress. This climate anxiety can be particularly prominent among young people, who are acutely aware that they will bear the brunt of climate impacts.

Of course, these impacts are not evenly distributed. As with most aspects of climate change, the mental health impacts disproportionately affect the most vulnerable: those with pre-existing mental health conditions, children and the elderly, socioeconomically disadvantaged individuals, and communities already bearing the brunt of environmental degradation.

In the face of these significant challenges, it's crucial to recognize that our mental health systems, as they currently stand, are ill-equipped to handle this impending crisis. To address this gap, a multifaceted approach is needed—one that combines adaptation strategies to minimize impacts and resilience-building measures to strengthen individual and community responses.

Mental health professionals have a crucial role to play in this respect. This includes not only providing therapeutic interventions for those directly affected by climate impacts but also advocating for recognition of these issues in broader climate policy discussions. More fundamentally, it involves helping to foster a societal paradigm shift—one that acknowledges our psychological interconnectedness with the planet and our mutual well-being.

Research also plays a critical role. Although our understanding of the mental health impacts of climate change has improved in recent years, there's still much to explore. Studies are needed to tease apart the intricate relationships between various climate stressors and mental health outcomes, and to identify the most effective interventions at individual, community, and policy levels.

Only by recognizing the profound psychological impacts of climate change can we hope to foster a future that is both physically and mentally resilient. As with all aspects of climate change, we need to act now—our mental health, and that of future generations, depends on it.

Adaptation measures are central in this effort. As climate change progresses, so too must our efforts to adjust our behaviors and systems in response to this new reality. This could mean incorporating climate change education into school curricula, not only to raise awareness about its environmental implications but also to address its mental health impacts.

Equally important is the role of resilience-building measures. Resilience, in this context, refers to the ability of individuals and communities to withstand, recover from, and adapt to climate-related stressors. This could take many forms, from community-led initiatives to foster social cohesion and mutual support, to therapeutic interventions to build personal coping skills and strategies. Supporting mental health in the face of climate change also requires targeted, evidence-based policies. Policymakers have a crucial role to play here, from ensuring that mental health services are adequately funded and accessible, to integrating mental health considerations into broader climate adaptation and resilience plans.

Further, there's a growing realization that nature itself can serve as a buffer against the mental health impacts of climate change. 'Green prescriptions'—where individuals are encouraged to spend time in natural settings—have been shown to have numerous mental health benefits. This highlights the potential value of urban green spaces, and of policies aimed at protecting and enhancing these spaces, for public mental health.

Now, let's turn our attention to the important role of research in addressing this crisis. While research in this area has grown in recent years, there remain substantial gaps in our understanding. More studies are needed to understand the varied and complex ways in which climate change impacts mental health across different populations and contexts. This includes research to identify the most vulnerable groups and to develop effective interventions for these groups.

There is also a pressing need for more longitudinal studies, which follow participants over extended periods, to understand the long-term mental health impacts of climate change. This includes not only the effects of acute climate-related events, but also the cumulative impacts of chronic environmental stressors.

Furthermore, research is needed to evaluate the effectiveness of various interventions aimed at mitigating the mental health impacts of climate change. This includes not only clinical interventions but also policy measures and community-based approaches.

The intersection of climate change and mental health is a pressing issue that demands urgent attention. With the world growing warmer, and with the mental health impacts of climate change becoming increasingly apparent, it's crucial that we take decisive action. This requires an integrated approach, incorporating adaptation strategies, resilience-building measures, targeted policies, and research efforts.

In this new reality, it's clear that the health of our planet and the health of our minds are inextricably linked. As we navigate the challenges of climate change, let us not forget that caring for our planet also means caring for our collective mental well-being. After all, it is not just the earth's resilience at stake—it's our resilience too.

Food Security and Nutrition: From Farm to Plate

In the face of climate change, it's imperative to remember that the implications go beyond the noticeable shifts in weather patterns and the accompanying natural disasters. Our agriculture, food systems, and thus our very plates are on the front lines.

One of the most stark examples of this comes when we examine food security and nutrition. The concept of food security extends beyond the mere presence or absence of food. It includes having access at all times to enough food for an active, healthy life for all household members, and also the variety and quality of that food. Climate change, with its repercussions on rainfall patterns, temperatures, and soil health, disrupts this intricate balance.

As climate change progresses, agricultural systems have been one of the first casualties. Changes in temperatures and precipitation patterns make it more difficult to grow crops in places where they have been grown for generations. For instance, coffee, a crop sensitive to temperature and rainfall, has already seen declining yields in parts of Central and South America. Such situations pose a dire threat to the livelihoods of millions of farmers worldwide.

Furthermore, the impact of climate change on the nutritional quality of food cannot be understated. Rising carbon dioxide levels are linked to reduced nutrient content in several staple crops, including wheat and rice. This nutrient dilution implies that even if people have enough to eat, they may still suffer from malnutrition due to the declining quality of their food.

The ripple effects are far-reaching, from the individual farmer to the global commodity markets, from rural economies to urban dinner tables, and from the fight against malnutrition to global health security.

In the face of such challenges, a shift is required in how we approach our food systems, from production to consumption. Sustainable agricultural practices need to be widely adopted to create resilient food systems that can withstand the impacts of climate change. This includes crop diversification, agroecology, agroforestry, and sustainable soil and water management techniques.

Farmers, who are at the front line of the climate crisis, need to be equipped with the necessary tools, training, and resources to adapt to changing weather patterns. Investment in agricultural research is key here, to develop and disseminate climate-smart crop varieties and farming practices.

Moving along the food chain, post-harvest practices also require attention. Significant amounts of food are lost post-harvest due to poor storage and transportation facilities. By addressing these gaps, we can ensure that a larger portion of what is produced actually makes it to our plates.

And then comes the important issue of consumption. Our diets play a huge role not only in our health but also in the health of our planet. Shifting towards diets that are both healthy and sustainable is an important part of the equation. This includes eating more plant-based foods and reducing food waste, which in itself accounts for about 8% of global greenhouse gas emissions.

It's also important to remember that the food security challenge isn't just about producing more food, but about making food systems more equitable. It's a grim paradox that many who produce the world's food are themselves food insecure. Policies that ensure fair income for smallholder farmers, equitable access to resources, and protection against market volatilities are crucial in this regard.

Lastly, while the food system contributes to climate change, it can also be part of the solution. Regenerative agriculture, for instance, can sequester carbon in soils, turning farms from carbon emitters to carbon sinks. Similarly, reducing food waste can significantly cut down on greenhouse gas emissions.

Climate change poses significant challenges to food security and nutrition, but it also opens opportunities for transformation. It's clear that we need a systemic change in our food systems—from farm to plate. By making our food systems more resilient, sustainable, and equitable, we can ensure that they can feed a growing population amidst a changing climate while contributing to climate mitigation. It's a tall order, but one that humanity cannot afford to shirk.

Resilience is a key concept that must be embraced in this journey. We need our food systems to be strong enough to weather the shocks and stresses brought about by climate change. This means developing agricultural systems that are diverse and flexible, capable of producing food under a range of conditions. It also means building supply chains that can quickly adapt to disruptions, and ensuring that households have the means to access nutritious food, even in times of crisis.

Sustainability, too, must be woven into every aspect of our food systems. This involves not only reducing the environmental footprint of food production, but also ensuring the social and economic sustainability of farming communities. Farmers need to be able to make a decent living from their work, and rural communities need to be vibrant and viable places for people to live.

Equity is the third cornerstone of the transformation we need. Today, access to food and the means to produce it are highly unequal, both within and among countries. Addressing these disparities requires concerted action at all levels, from local to global.

As we strive to make our food systems more resilient, sustainable, and equitable, we also have to recognize the crucial role of public policy. Governments have a key role to play in creating enabling environments for sustainable agriculture, in promoting healthy diets, and in protecting the most vulnerable from food insecurity. Indeed, the transformation of our food systems is not only a technical challenge but also a political one.

Furthermore, the task is not one that governments can accomplish alone. All actors in the food system have a part to play, from farmers to consumers, and from businesses to NGOs. It's only through collective action that we can bring about the systemic changes we need.

In all this, science and technology have crucial roles to play. From climate-smart crop varieties to innovations in food storage and processing, scientific research can provide many of the tools we need to make our food systems more climate-resilient. But these tools need to be made accessible to all, especially to smallholder farmers in developing countries, who are often the most vulnerable to climate change.

At the same time, traditional knowledge should not be overlooked. Indigenous and local communities have a wealth of knowledge on how to live in harmony with nature and adapt to changing environments. Their insights can complement scientific knowledge and provide valuable guidance in our quest for sustainable food systems.

In sum, the challenges posed by climate change to food security and nutrition are daunting, but they are not insurmountable. By embracing resilience, sustainability, and equity, and by harnessing the power of both science and traditional knowledge, we can transform our food systems and ensure that everyone has access to the nutritious food they need.

It's a challenge we must meet, not only for the sake of our own generation but also for those to come. As we navigate through the age of adaptation, let's make sure that no one is left behind.

Chapter 11: The Role of Climate Change in the Anthropocene

Defining the Anthropocene: Man's Geological Epoch

The term "Anthropocene" is etched in the narrative of climate change, a word spoken in hushed reverence, indicative of the tremendous influence of humanity on the Earth's systems. Its inception marks the advent of an epoch where human activities have begun to have a significant global impact on the Earth's ecosystems. It is the geologic chronological term proposed to mark the evidence and extent of human activities that have had a substantial global influence on the Earth's ecosystems. But what exactly does this term entail? What are its implications, and why is it relevant in the context of climate change?

The term Anthropocene, originating from the Greek words 'anthropos' (human) and 'kainos' (new), was first proposed by ecologist Eugene F. Stoermer and popularized by Nobel Laureate Paul Crutzen in the early 2000s. This era, still unofficial in the geological time scale, is hypothesized to have begun in the late 18th century with the Industrial Revolution. This was when our species first started to alter the composition of the atmosphere by pumping out vast amounts of carbon dioxide from the burning of fossil fuels, a trend that has continued and accelerated to this day.

Yet, defining the Anthropocene is more than just marking the timeline of human industrial activity. It acknowledges a fundamental shift in the relationship between humans and the Earth, a shift where we have moved from being passive observers of nature to becoming active, even dominant, players in the planetary system.

The Anthropocene epoch represents a new phase in the history of both humankind and of the Earth, when natural forces and human forces became intertwined, so much so that the fate of one determines the fate of the other. Geologically, this is a period marked by the radioactive elements dispersed across the planet by nuclear bomb tests, although many other stratigraphic signals may also reflect the human impact upon the global environment.

The climate change crisis we face today is the most palpable manifestation of the Anthropocene. The unbridled burning of fossil fuels, deforestation at unprecedented scales, and the rise of industrial farming are just some of the human-induced changes that have disrupted the Earth's natural carbon cycle and led to the current warming trend.

Recognizing our epoch as the Anthropocene also holds profound implications for the way we perceive our role and responsibilities towards the Earth. It underscores the reality that we are now the stewards of our planet, whether we like it or not, and that our collective actions will determine the trajectory of the Earth system over millennia.

The Anthropocene presents a paradoxical notion of power and vulnerability. On the one hand, it reveals the immense power humans hold to shape the future of the Earth system. On the other hand, it also exposes our vulnerability, as the environmental changes we trigger could ultimately threaten our own survival.

Thus, defining the Anthropocene isn't merely an academic exercise, but also an urgent call for action. It's a recognition of the magnitude of our impact on the Earth – a testament to our potential to destroy but also, perhaps, our ability to protect and preserve.

In the age of the Anthropocene, mitigating climate change and adapting to its impacts aren't just environmental or economic challenges. They're also ethical and existential ones. They compel us to rethink our values, our societal structures, and our relationship with the natural world.

As we navigate the Anthropocene, we must strive to become not just wise stewards but also humble students of our planet. We must learn from the Earth's long history, understand its complex systems and interdependencies, and respect its limits. At the same time, we must harness our collective knowledge, technology, and spirit of innovation to design a future where humans and nature can thrive together.

In the end, the Anthropocene might be the impetus we need to transition towards a more sustainable and resilient civilization. It highlights the urgency of adopting a new worldview that recognizes the interconnectedness of all life on Earth and respects the intricate balance of its natural systems.

The Anthropocene has a dual role in the context of climate change. On one hand, it is a problem statement, a concept that encapsulates the adverse global impacts of human activities. On the other hand, it is also a call to action—a prompt for us to leverage our vast capabilities and resources for the betterment of the planet.

The Anthropocene is a mirror, reflecting the massive changes we have made to the Earth, changes that have caused irrevocable harm and accelerated climate change. However, if we change our perspective, the mirror can also serve as a window—a view into a new world that we can shape using the same tools and technologies that caused harm in the first place.

The Anthropocene era emphasizes human influence on climate, which means that just as we have the power to destroy, we also have the power to preserve and protect. This power can be harnessed through numerous ways: reducing greenhouse gas emissions, increasing energy efficiency, fostering sustainable agricultural practices, protecting and restoring ecosystems, and transitioning to a circular economy.

Recognizing our current epoch as the Anthropocene also means acknowledging our responsibility to future generations. It is our duty to ensure they inherit a world where they can live and thrive, a world where nature is not just a resource to be exploited but a complex, intertwined system to be respected and preserved. This is the mindset that we need to cultivate—an Anthropocene mindset.

We can no longer afford to see ourselves as separate from the natural world. The traditional dichotomy of humans versus nature is not just outdated, but also harmful. In the Anthropocene, we are one with nature. We are the drivers of change, and with that, comes responsibility.

As we stride into the Anthropocene, we face a choice. We can continue along our current path, contributing to environmental degradation and exacerbating climate change. Or, we can embrace the significance of this new epoch, accept our roles as stewards of the planet, and work tirelessly towards creating a more sustainable, equitable, and resilient world.

Each one of us has a role to play. From policy makers and business leaders to scientists, teachers, and individuals, we all have the capacity to make a difference. Whether it's through implementing climate policies, investing in green technologies, conducting groundbreaking research, educating the next generation, or making sustainable lifestyle choices, we can all contribute to mitigating the impacts of climate change.

The Anthropocene epoch presents us with challenges of an unprecedented scale and complexity. But it is also a testament to our potential—a reminder of the remarkable power we hold to shape the Earth and its future. In this epoch, we are not just witnesses to change but also the architects of it. The challenge of the Anthropocene, then, is not just about surviving but about thriving—not just about adapting to change but about driving it.

In the face of this challenge, there's no room for despair. The Anthropocene calls for action, resilience, and most importantly, hope. It's a call that we must heed for the sake of our planet and our future. The Age of Adaptation, thus, isn't just about adapting to a changing world—it's also about changing the world for the better.

Climate Change: The Major Driver of the Anthropocene

Indeed, there is no doubt that climate change is the major driver of the Anthropocene era, the epoch where human activities began to have a substantial global impact on Earth's ecosystems. It is the clearest and most apparent sign of our growing influence on the natural world, with ramifications that extend far beyond the scientific and environmental realms, encompassing social, economic, and cultural aspects of our lives.

Climate change, driven largely by the burning of fossil fuels, deforestation, and other forms of land-use change, has already caused global temperatures to rise by about 1.2°C since the pre-industrial era. If current trends continue, we are likely to see a warming of 3-4°C by the end of this century—a level of warming that could render large parts of the Earth uninhabitable, disrupt critical ecosystems, and pose existential risks to humanity and countless other species.

In addition to raising global temperatures, climate change is also altering precipitation patterns, leading to more intense and frequent extreme weather events, from floods and hurricanes to droughts and heatwaves. These changes are triggering a cascade of secondary effects, including sea-level rise, ocean acidification, the melting of glaciers and ice sheets, and shifts in biodiversity and ecosystem services. In short, climate change is not only a driver of the Anthropocene—it's a force that's reshaping the face of the Earth.

However, in the context of the Anthropocene, climate change isn't just an environmental issue—it's a symptom of a much deeper problem. It's a mirror reflecting our dysfunctional relationship with the natural world, a consequence of viewing the Earth as a commodity, and a sign of the unsustainable trajectory of our current economic and social systems. It's a manifestation of the Anthropocene's key paradox: our remarkable capacity for innovation and adaptation, which has brought unprecedented prosperity and progress, is also driving the destabilization of the Earth system upon which we depend. In response to this crisis, humanity needs to rethink its relationship with the planet. The Anthropocene demands a new paradigm—a shift from exploitation to stewardship, from short-term gain to long-term sustainability, from a human-centric worldview to one that recognizes the intrinsic value of all life and the interconnectedness of all systems on Earth.

One of the critical lessons of the Anthropocene is that we cannot solve climate change in isolation. It's intertwined with other global challenges, such as biodiversity loss, pollution, social inequality, and unsustainable consumption and production patterns. Addressing these problems requires a holistic, integrated approach—one that goes beyond reducing greenhouse gas emissions to transforming our economic, political, and social systems.

We also need to harness the power of science, technology, and innovation to create a more sustainable and resilient world. In the Anthropocene, these tools are not just optional—they're essential. They can help us develop cleaner sources of energy, create more efficient and sustainable production systems, protect and restore critical ecosystems, and build resilient communities. At the same time, we must recognize that these tools are not silver bullets—they are part of a broader toolkit that includes political will, social change, and individual action.

Ultimately, the Anthropocene forces us to confront a fundamental question: What kind of world do we want to live in? This question, which goes to the heart of our values, aspirations, and responsibilities, cannot be answered by science alone—it's a question for all of humanity.

The Anthropocene is an epoch of profound challenges, but it's also an epoch of extraordinary opportunities. It's an era that forces us to reckon with the consequences of our actions, to rethink our relationship with the natural world, and to reimagine our future. In this sense, climate change is not just a driver of the Anthropocene—it's a catalyst for change, a wake-up call that compels us to create a more sustainable, equitable, and resilient world.

In particular, the Anthropocene challenges us to view climate change not just as a threat, but as an opportunity for transformation. An opportunity to shift away from fossil fuels and towards renewable energy. An opportunity to move away from resource-intensive consumption and towards sustainable lifestyles. An opportunity to transition from exploitative economies and towards models of circular, regenerative economies. And above all, an opportunity to recognize the fundamental interconnectedness of all life on Earth and to strive for a world where humans live in harmony with nature rather than dominate it.

Of course, seizing these opportunities won't be easy. It requires bold leadership, forward-thinking policies, and collective action at all levels, from individual citizens and communities to businesses and governments. But if there's one thing the Anthropocene teaches us, it's that we have the capacity to shape our own future. After all, we are the architects of this epoch.

The Anthropocene also underscores the importance of education and awareness in addressing climate change. As the major driver of the Anthropocene, climate change needs to be understood by all—not just scientists and policymakers but every individual across the globe. Climate education is a powerful tool for empowering people to take action and make informed decisions, helping us navigate the complexities and uncertainties of this new epoch.

Moreover, as we confront the challenges of the Anthropocene, we must not forget the power of human resilience and adaptability. These qualities have enabled us to survive and thrive in diverse environments across the globe, from the frigid Arctic to the scorching Sahara. In the face of climate change, these qualities will be more crucial than ever, helping us adapt to new conditions, innovate solutions, and build a resilient future.

In conclusion, the Anthropocene—an epoch defined by human influence—is a potent symbol of our impact on the planet. Climate change, as the major driver of the Anthropocene, represents the most visible and far-reaching manifestation of this influence. But it also offers us an unprecedented opportunity for change.

The era is, undoubtedly, fraught with challenges. Yet, it also provides an impetus for us to reimagine our relationship with the natural world and redefine our place within it. As we journey through the Anthropocene, our ability to drive positive change, champion sustainability, and cultivate resilience will not only determine our response to climate change—it will define the legacy of our species in this unique epoch.

In the grand narrative of Earth's history, the Anthropocene will be but a single chapter. But it is our chapter. And the story it tells—of challenge and change, of adaptation and resilience—will echo far into the future, shaping the trajectory of life on Earth for millennia to come. So, let us write this chapter not with a sense of despair, but with a spirit of hope and determination, as we work together to address the challenges of climate change and strive for a sustainable and resilient future for all.

Adaptation Strategies for Thriving in the Anthropocene

As we navigate through the complexities of the Anthropocene, the epoch defined by our own human influence, we are faced with a profound realization: the urgency of adapting to the rapid environmental changes driven primarily by climate change. But how can we thrive in the Anthropocene? What adaptation strategies can we deploy to not only survive but prosper in these changing times?

Firstly, the Anthropocene calls for transformative change, the type of change that goes beyond incremental adjustments and aims at altering the very systems that have propelled us into this new epoch. This includes a paradigm shift in our economic systems, moving from linear, growth-obsessed models to regenerative and circular economies that value sustainability over short-term gain. Such transformation requires innovative business models, policy incentives, and a new consciousness among consumers about the environmental impacts of their consumption habits.

Examples of circular economies are already being implemented worldwide. In the Netherlands, the city of Rotterdam is redesigning its port to become a hub for circular industries, recycling waste, and using bio-based materials. Meanwhile, companies like Patagonia are pioneering business models that prioritize durability, repairability, and recycling of their products.

Secondly, adapting to the Anthropocene necessitates a reimagining of our energy systems. Fossil fuels, the primary driver of anthropogenic climate change, must be phased out in favor of renewable energy sources like solar, wind, and hydro power. This transition not only mitigates climate change but also brings about opportunities for sustainable development and green jobs.

Countries like Denmark and Costa Rica are leading the way in renewable energy transition. Denmark, powered by robust wind energy, aims to be independent of fossil fuels by 2050, while Costa Rica already generates more than 98% of its electricity from renewables. These successes demonstrate that a renewable energy future is not just possible—it's already happening.

Thirdly, thriving in the Anthropocene requires a revolution in our food systems. With agriculture contributing to nearly a quarter of global greenhouse gas emissions and being highly vulnerable to climate change, sustainable farming practices are crucial. This means promoting agroecology, organic farming, and permaculture, which enhance biodiversity, improve soil health, and sequester carbon. Moreover, we need to reduce food waste, which accounts for about 8% of global emissions, and shift towards more plant-based diets.

There are inspiring examples of sustainable agriculture from around the world, from the terraced rice fields of Bali that have sustained communities for centuries, to the urban farms sprouting in cities like Detroit and Havana, turning vacant lots into productive green spaces. These initiatives show that sustainable agriculture can not only feed us but also heal the land and the climate.

Beyond these systemic changes, adaptation in the Anthropocene must also be rooted in social equity and justice. As the impacts of climate change disproportionately affect the most vulnerable, adaptation strategies must prioritize the needs of these communities. This means investing in climate-resilient infrastructure in vulnerable regions, ensuring equitable access to resources like clean water and energy, and empowering communities, particularly women and Indigenous peoples, in decision-making processes.

Initiatives like the Slum Dwellers International, which works to improve living conditions in informal settlements, and the Indigenous-led conservation efforts, which protect biodiversity and sequester carbon while safeguarding cultural heritage, exemplify the power of equitable, community-led adaptation.

Finally, adapting to the Anthropocene involves fostering psychological resilience. The environmental changes and extreme weather events we're witnessing can cause significant psychological stress, leading to a rise in mental health issues. Promoting mental health resilience involves increasing awareness about these issues, providing psychological support to affected communities, and building social connections that can buffer against psychological stress.

In Australia, where extreme droughts and wildfires have taken a mental toll on farmers and rural communities, initiatives like the Rural Adversity Mental Health program have been established to provide support and resources to these communities. The program not only offers direct mental health services, but also engages in community outreach to raise awareness about mental health issues and break down the stigma associated with seeking help.

The potential for education in fostering resilience should also not be overlooked. Climate change education that goes beyond simply communicating the science and also addresses the social and emotional aspects of the crisis can be a powerful tool for building resilience. Such education can cultivate a sense of agency and hope, equip people with the skills to address climate challenges, and foster a deep sense of connection to the Earth and to each other.

Organizations like Climate Generation offer a model for this type of holistic climate change education. Their curriculum, designed for both students and educators, includes not only climate science, but also the sociopolitical aspects of climate change, personal and community resilience strategies, and opportunities for action and advocacy.

Adapting to life in the Anthropocene also calls for novel ways of relating to nature. As we recognize that we are not separate from the environment, but deeply embedded within it, we can begin to cultivate a sense of care and stewardship for the natural world. This might involve embracing traditional ecological knowledge, which often embodies a deep respect for nature, or fostering a sense of biophilia, an innate love for nature, through spending time in natural settings.

For example, in Aotearoa New Zealand, the Whanganui River has been granted the same legal rights as a human being, reflecting the Maori worldview that sees humans as deeply interconnected with the natural world. Similarly, the Forest School movement, which originated in Scandinavia and has spread globally, encourages children to learn in outdoor environments, fostering a sense of curiosity, wonder, and respect for nature.

The time we are in is undeniably a challenging epoch, characterized by rapid environmental change and unprecedented uncertainties. However, it also presents opportunities for creativity, innovation, and transformation. The adaptation strategies discussed here—from transformative economic systems and renewable energy transitions to sustainable food systems, social justice, mental resilience, education, and a new relationship with nature—are not just necessary for survival, but for thriving in the Anthropocene.

While these strategies are diverse, they are all interconnected and reinforce each other, creating a virtuous cycle of change. Each of these elements alone cannot solve the climate crisis, but together, they provide a blueprint for a resilient and prosperous future in the Anthropocene.

It's not going to be easy, and it will require cooperation on a global scale and at all levels of society. But as we've seen from countless examples worldwide, humans are remarkably adaptable and creative creatures. And with that adaptability and creativity, we can chart a new path forward in the Anthropocene, transforming crisis into opportunity and forging a future that is not only sustainable but also just, resilient, and deeply fulfilling.

This is the task of our generation. It's a big one, but it's also an incredible opportunity. The age of adaptation in the Anthropocene is not just about survival—it's about thriving in a changed world.

As we embark on this journey, we are not only adapting to a changing climate—we are reshaping our world and our minds, building a future that honors both people and the planet. And in doing so, we are redefining what it means to be human in the Anthropocene.

Chapter 12: Futuristic Vision: Imagining a Climate-Resilient World

The Role of Innovation and Technology in Climate Adaptation

As we delve into the final chapter of our book, we venture into the world of possibilities and envision the role of innovation and technology in climate adaptation, the foundation of our futuristic vision for a climate-resilient world. But before we do so, let's remind ourselves of one fact - adaptation is not an option, it's a necessity. As we navigate through the Anthropocene, our survival and prosperity hinge on our ability to adapt to the rapidly changing climatic conditions. Innovation and technology play pivotal roles in facilitating that adaptation, offering solutions that can mitigate the impacts of climate change and help societies to become more resilient.

We are in the midst of a technological revolution that is reshaping every facet of our lives, from the ways we communicate, to the ways we work, travel, consume, and even think. As formidable as the challenges of climate change are, it's worth remembering that we have more tools at our disposal than at any other time in human history. Innovation and technology hold the potential to drastically reduce our greenhouse gas emissions, enhance our resilience to climate change impacts, and transform the way we live and interact with our environment.

When we think about climate innovation, one of the first things that likely comes to mind is renewable energy. Indeed, the transition from fossil fuels to renewable energy sources is one of the most critical and necessary steps towards mitigating climate change. Technological advancements in this field, such as improvements in solar panel efficiency, wind turbine design, and battery storage capacity, have made renewable energy more affordable and accessible than ever before.

Take, for example, the significant improvements in energy storage technology. Tesla's development of the Powerwall, a rechargeable home battery system that stores solar energy for use during nighttime or power outages, exemplifies the potential for innovation to transform our energy systems. This technology not only promotes the use of clean energy but also increases resilience by offering a reliable power source during extreme weather events, which are likely to increase due to climate change.

But the domain of climate innovation extends far beyond renewable energy. In the realm of agriculture, for instance, precision farming technologies, such as GPS-guided tractors and drones for crop monitoring, are enabling farmers to manage their resources more efficiently, reducing water and fertilizer use, and increasing crop yields. This is a crucial adaptation strategy in a world where food security is increasingly threatened by climate change.

Similarly, advancements in biotechnology are opening up new possibilities for climate-resilient crops. Scientists are now able to engineer crops that can withstand higher temperatures, drought, and even salinity, offering hope for food production in regions that are most vulnerable to climate change.

Even the built environment is being reimagined through the lens of climate resilience. We see innovative designs for climate-resilient infrastructure, ranging from flood-proof houses in the Netherlands, to heat-resistant roads in Los Angeles, to sponge cities in China designed to absorb heavy rainfall and prevent urban flooding.

Emerging technologies such as artificial intelligence (AI) and machine learning are also promising tools for climate adaptation. These technologies can be used to model and predict climate change impacts, optimize renewable energy production, and even monitor deforestation in real-time.

Moreover, AI could assist in anticipating climate-induced health crises. For example, machine learning algorithms can predict outbreaks of vector-borne diseases, such as malaria and dengue, based on weather patterns, allowing for proactive public health interventions.

Of course, we must also acknowledge the potential risks and challenges associated with these technological solutions. The use of AI and big data raises concerns about privacy and data security. Similarly, bioengineered crops might have unintended consequences for biodiversity.

Technological solutions are not a panacea, and they should not be pursued in isolation from other adaptation strategies, such as ecosystem restoration, sustainable land management, and the promotion of social equity. However, combined with these strategies, technology and innovation have the potential to significantly bolster our resilience to climate change.

The promise of technology is not just about developing new tools for adaptation, but also about harnessing and applying existing technologies in innovative ways. Take, for instance, the humble mobile phone. In regions prone to extreme weather events, simple text message alerts can save lives by warning people of impending storms or floods. Mobile banking and microinsurance services can help communities to recover more quickly from climate-related disasters. And mobile apps can connect farmers with up-to-date information about weather forecasts, crop prices, and sustainable farming techniques. All these examples underscore that sometimes the most effective solutions are not the most high-tech or expensive ones.

Even in the face of a changing climate, technological advancements offer the potential to not just survive, but thrive. They can create jobs, improve health and wellbeing, and even catalyze social and economic development. The German town of Feldheim, for instance, has become energy independent through local wind and solar power generation, creating jobs and attracting tourists in the process.

What becomes clear is that innovation and technology can empower communities to take charge of their climate future. Climate adaptation, then, is not just about building physical resilience, but also about fostering adaptive capacity – the ability to anticipate, respond to, and learn from change.

It is important to note, however, that the benefits of technological advancements are not evenly distributed. Access to technology is often limited in the very regions that are most vulnerable to climate change. Bridging this "climate technology gap" is an urgent priority. This requires not only investment in infrastructure and education but also policies that promote technology transfer and cooperation on a global scale.

We can imagine a future where we harness the power of technology to build a climate-resilient world. A world where renewable energy powers our cities, precision farming feeds our population, AI helps us anticipate and manage climate risks, and every individual has access to the technology they need to adapt to a changing climate.

The challenges we face are monumental, but so too are the opportunities. As we stand at the precipice of the

Anthropocene, our actions in the coming years will shape the future of our planet and our species. We have the knowledge, the tools, and the innovation to turn the tide against climate change. We must now have the courage to use them, and the wisdom to use them wisely.

Our journey through the Age of Adaptation has shown us that climate change is not just a crisis, but also a catalyst for change. As we confront the realities of our warming world, we are being challenged to rethink and reinvent our ways of living, working, and coexisting with the natural world. The future may be uncertain, but one thing is clear: adaptation is our path forward, and innovation and technology will light the way.

Designing Climate-Resilient Cities of the Future

Designing climate-resilient cities of the future requires a holistic and integrated approach. This is about much more than just engineering and architecture. It's about creating urban environments that are sustainable, adaptable, and inclusive, where communities can thrive in the face of climate change.

Our cities are particularly vulnerable to the impacts of climate change. Rising sea levels threaten coastal cities, heatwaves pose risks to health and infrastructure, and extreme weather events can lead to devastating floods. Furthermore, the urban heat island effect - where cities are significantly warmer than their surrounding rural areas due to human activities - exacerbates these challenges.

However, cities are not only sites of vulnerability but also hubs of innovation. In the crucible of the urban environment, with its concentration of people, resources, and ideas, we have a unique opportunity to forge new pathways towards resilience.

The city of the future must be designed with resilience at its core. This means creating urban environments that can absorb, recover from, and adapt to climate impacts. It means designing buildings, infrastructure, and public spaces that are flexible and robust, capable of withstanding the pressures of a changing climate.

Crucially, the resilient city is also a green city. Green spaces, parks, and urban forests can reduce the urban heat island effect, absorb stormwater to prevent flooding, and enhance biodiversity. They can also provide spaces for recreation and social interaction, contributing to the wellbeing of urban dwellers.

Take, for example, the city of Melbourne in Australia. Melbourne's Urban Forest Strategy aims to increase canopy cover to 40% by 2040 to combat the urban heat island effect and to create a more livable city. The city uses a digital tool called Urban Forest Visual to track the health, age, and species of every public tree in the city, inviting citizens to "email a tree" to report problems or simply express their appreciation.

Furthermore, the resilient city is a smart city, harnessing technology to optimize resource use, reduce emissions, and improve the quality of urban life. Smart grid technologies can balance electricity supply and demand, reducing the risk of blackouts during heatwaves. Internet of Things (IoT) devices can monitor environmental conditions in real-time, providing crucial data to manage climate risks. And digital platforms can engage citizens in urban planning, fostering a sense of ownership and community.

Cities also need to be designed with social equity in mind. Climate change disproportionately affects disadvantaged communities, and without careful planning, climate adaptation measures can exacerbate social inequalities. Resilient urban design must therefore aim to create not only safer but also fairer cities. This means ensuring that all citizens, regardless of their socio-economic status, have access to green spaces, quality housing, and resilient infrastructure. It means involving communities in decision-making processes, respecting their rights, and responding to their needs.

The city of Medellin in Colombia provides an inspiring example. In the face of high temperatures and poverty, the city embarked on an ambitious project to build green corridors and a cable car system to connect marginalized hillside neighborhoods with the city center. The project has reduced temperatures, improved air quality, and fostered social inclusion.

Lastly, the resilient city is a regenerative city. It's not enough to simply withstand climate impacts; we need to create urban systems that restore and regenerate the natural environment. This means moving towards circular economy models, where waste is minimized and resources are continually reused. It means integrating nature into the urban fabric, creating living buildings with green roofs and walls, and revitalizing waterways.

The Danish city of Copenhagen exemplifies this vision. Its CopenHill project transformed a waste-to-energy power plant into a multi-purpose facility featuring an artificial ski slope, hiking trails, and a climbing wall. The project showcases the potential of integrating urban design, waste management, and recreation to create spaces that are both functional and enjoyable.

Designing climate-resilient cities also involves proactive planning for sea-level rise and coastal erosion. Cities like New York and Amsterdam are leading the way in this regard. New York's Big U project, a proposed protective system around the lower half of Manhattan, includes elevated parks and spaces for markets and social events, showing that resilience infrastructure can also serve as a communal space. Similarly, Amsterdam is experimenting with floating neighbourhoods to adapt to rising sea levels, such as the IJburg district, which houses thousands of residents on artificial islands.

Transportation is another key aspect. Reducing reliance on personal vehicles decreases greenhouse gas emissions and improves air quality. Future cities will increasingly prioritize active modes of transport like walking and cycling, along with public transit. Additionally, as electric vehicles become more widespread, cities will need to adapt by creating sufficient charging infrastructure. City planning that encourages mixed-use development reduces the need for long commutes, further reducing a city's carbon footprint.

But, urban climate resilience is not just about the physical design of cities; it's also about community and culture. The climate-resilient city of the future fosters a culture of sustainability and inclusivity. This includes education about climate change and its effects, promoting sustainable practices like recycling and energy efficiency, and actively involving residents in decisions that affect their environment and community.

We also need to recognize that every city is unique, with its own set of climate risks, resources, and cultural context. This means that there is no one-size-fits-all solution to climate resilience. Each city must craft its own resilience strategy, tailored to its particular needs and circumstances.

The climate-resilient cities of the future will be much more than just physical spaces; they will be vibrant, inclusive communities where sustainability is woven into the fabric of daily life. They will be places where humans and nature coexist in harmony, and where the challenges of climate change are met with innovation, resilience, and a shared sense of purpose.

Climate resilience is not a distant goal, but an urgent necessity. And by reimagining our cities, we have a chance to create a future that is not only resilient but also healthier, fairer, and more sustainable. The task is daunting, but with creativity, collaboration, and commitment, it is within our reach. As we move forward into this uncertain future, let us seize the opportunity to redefine what a city can be, to create urban spaces that are not only adapted to climate change, but are also thriving, vibrant communities that enrich our shared world.

Envisioning a New Social Contract: Climate Justice and Equality in the Anthropocene

The Anthropocene, characterized by the unprecedented influence of human activities on the Earth's systems, presents new challenges that demand a radical rethinking of our social contract. It requires us to look beyond the short-term and consider the long-term effects of our actions on the planet and future generations. Climate justice and equality become central pillars of this new social contract, ensuring that the burdens and benefits of climate adaptation are shared equitably, and the voices of all, especially the most vulnerable, are heard and valued in decision-making processes.

Climate justice is a term that captures the idea that those who have contributed least to climate change, often suffer the most from its impacts. It acknowledges the uneven distribution of climate change effects and seeks to address these inequities. In the Anthropocene, climate justice necessitates a social contract that places a greater responsibility on high-emitting nations and corporations to significantly cut emissions and contribute more towards climate adaptation and mitigation efforts globally. This requires a dramatic shift in our economic and social systems, moving away from a purely profit-driven model towards one that prioritizes ecological sustainability and social equity.

This new social contract must also be rooted in the principles of intergenerational equity, considering the needs of future generations. This means redefining progress not just in terms of economic growth, but also in terms of environmental sustainability and resilience. For instance, countries must rethink how they measure success, moving beyond GDP to other measures of well-being and sustainability, such as the Genuine Progress Indicator (GPI) or the Happy Planet Index.

An essential part of this new social contract is acknowledging and rectifying environmental racism. Communities of color and Indigenous populations around the world have been disproportionately affected by environmental degradation and climate change due to systemic racism and socioeconomic inequalities. They are often located in areas with higher pollution levels, have fewer resources to adapt to climate impacts, and their traditional lands and lifestyles are threatened. The new social contract needs to ensure that these communities are given equal protection, have access to resources for climate adaptation, and their rights to land, culture, and self-determination are respected.

On a more local level, this new social contract could take the form of inclusive urban planning that ensures all citizens have access to green spaces, clean air, and are protected from climate risks. It could also involve policies that protect workers in industries affected by the transition to a low-carbon economy, such as retraining programs and social safety nets.

Moreover, the social contract in the Anthropocene must address gender inequalities in climate impacts. Women and girls in many parts of the world bear the brunt of climate change due to existing gender inequalities. They often have fewer resources to adapt and are more dependent on natural resources for their livelihoods. Addressing gender inequality is not just a moral imperative; it is also crucial for effective climate adaptation. Policies and initiatives that empower women and girls, including access to education, healthcare, and economic opportunities, must be a key part of the new social contract.

Building a new social contract in the Anthropocene also demands a shift in our relationship with nature. We need to move from a view of nature as a resource to exploit, to one of a partner with which we share our planet. This requires recognizing the inherent value of biodiversity and ecosystems, not just their economic value, and adopting practices that respect and protect them. Indigenous knowledge systems that have long held such a perspective can provide valuable insights in this regard.

The Anthropocene demands a new social contract, one that is rooted in climate justice, equality, intergenerational equity, and a renewed partnership with nature. Such a contract is not only a response to the challenges of the Anthropocene but also an opportunity to create a more equitable, sustainable, and resilient world.

Final Thoughts on Climate Change and The Age of Adaption

As we reach the end of our exploration of climate change and the Age of Adaption, we find ourselves standing at a crucial juncture in the timeline of our species and the planet. The path that we've walked has been marked by dramatic shifts and changes, innovations and regressions, and above all, by the ever-evolving relationship between humans and our environment. The crisis we're confronting now—the climate crisis—is a direct result of that relationship. Yet within this crisis also lies the seeds of opportunity. The opportunity to adapt, to evolve, and to chart a new course for our shared future.

Throughout this book, we've delved into the science, politics, psychology, and sociology of climate change. We've touched on the experiences of vulnerable communities, the rich insights provided by indigenous knowledge systems, and the role of creative mediums in bringing home the realities of climate change. We've highlighted the urgency and necessity for both individual and collective action, and underscored the need for comprehensive policy changes.

Now, it's time to look forward, to cast our gaze towards the horizon and imagine the world that could be. Climate change, undoubtedly, is going to test us. It will test our resilience, our adaptability, and our compassion. But it will also force us to innovate, to reimagine our systems, and to redefine our relationship with the natural world. We've stepped into the Anthropocene, an epoch where human actions have become the driving force shaping our planet. But let's remember that this influence can also be wielded for positive transformation.

When it comes to climate change, we're often caught between two contrasting narratives: the doomsday scenario and the techno-optimistic future. The truth, as always, likely lies somewhere in the middle. There's no denying the magnitude of the challenges we face, but there's also no denying our potential to overcome them. We must embrace this complexity as we move towards creating solutions. The Age of Adaption is not about a singular grand solution to climate change. Instead, it's about embracing a mosaic of approaches that reflect the diverse needs and realities of our global community.

As we take steps into this future, our guiding principles must be grounded in justice, equity, and sustainability. Climate justice is not just a moral imperative—it is a practical necessity for creating resilient societies. We must ensure that adaptation strategies are inclusive, prioritizing the needs of those most vulnerable to the impacts of climate change.

We must also build new models of economic growth that respect planetary boundaries and prioritize well-being over consumption. The Anthropocene calls for a reimagining of our measures of progress, shifting from growth-at-all-costs to a model that balances human development with ecological integrity.

Let's not forget the critical role that innovation and technology will play in our journey towards adaptation. From renewable energy systems and carbon capture technologies to climate-resilient infrastructure and digital tools for enhancing climate literacy, the possibilities are truly endless. Yet, as we innovate, let's also remember to respect and learn from the wisdom of traditional knowledge systems that have long nurtured a harmonious relationship with nature.

Finally, we must remember that adaptation is not just about surviving—it's about thriving. It's about envisioning and creating a world that is not just resilient in the face of climate change, but also more equitable, more compassionate, and more in tune with the rhythms of the natural world.

The Age of Adaption is a call to action. It is a call to all of us, regardless of our background, profession, or geographic location, to become active participants in shaping our shared future. It is a call to learn, to care, and above all, to act.

As we close this chapter of our exploration, let's not see it as the end of the conversation, but as the beginning of a new journey—one filled with challenge, but also hope.
Let's harness the power of human ingenuity and the strength of collective action to build a future where we do not just survive climate change, but flourish in spite of it. This isn't just about saving the planet—it's about securing our future as a species, ensuring a world where future generations can thrive, and nurturing a harmonious relationship between humans and the environment.

As we step into this future, we carry with us not just the lessons of the past and the knowledge of the present, but also the dreams of what could be. The Age of Adaption is here. It's a challenge, a responsibility, but above all, an opportunity. And it's up to each of us to seize it.

This book may be ending, but our shared journey towards a climate-resilient future is just beginning. The narrative of climate change and our responses to it is being written right now, and every single one of us has a part to play in shaping it. So, let's write it well.

As we grapple with the hard truths of climate change and the Anthropocene, let's also allow ourselves to be inspired by the possibilities of what we can achieve together. Yes, we're faced with a monumental challenge, but we also have the opportunity to turn this challenge into one of the greatest stories of adaptation, resilience, and transformation the world has ever seen.

So, as we step into the Age of Adaption, let's do so with determination, with courage, and with hope. Because if there's one thing that's become clear throughout the course of this book, it's this: the future is not something that simply happens to us. It's something that we create. And in this age of unprecedented change, we have the opportunity to create something truly extraordinary.

Thank you for joining me on this journey through the Age of Adaption. I hope that it has provided you with valuable insights, sparked meaningful conversations, and inspired you to action. Because if there's one thing I hope you take away from this book, it's this: in the face of climate change, we are not powerless. We have the tools, the knowledge, and the capacity for innovation that we need to adapt and thrive.

The path ahead may be uncertain, but it's also ripe with possibility. And as we venture into the unknown, remember: we're not just adapting to survive. We're adapting to thrive. And that makes all the difference.

So here's to the Age of Adaption. Here's to the future we're going to build together. And here's to each and every one of us—because the power to shape the future is, and always has been, in our hands. Let's use it wisely.

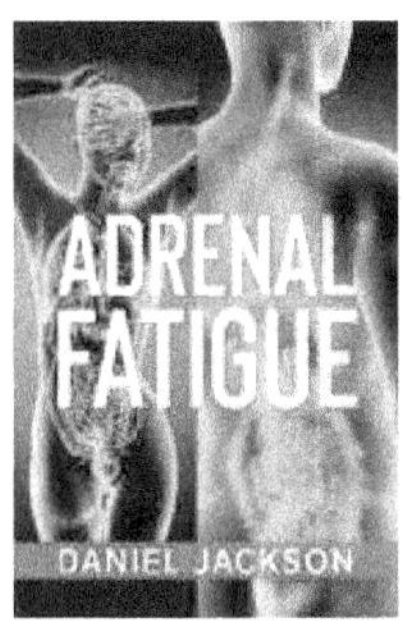

Take a look at more great books available from Rockwood Publishing

... some for FREE!

Just visit the link below:

rockwoodpublishing.co.uk

to the subject matter covered. The information included in this book has been compiled to give an overview of the subject(s) and detail some of the symptoms, treatments etc. that are available to people with this condition. It is not intended to give medical advice. For a firm diagnosis of your condition, and for a treatment plan suitable for you, you should consult your doctor or consultant. The writer of this book and the publisher are not responsible for any damages or negative consequences following any of the treatments or methods highlighted in this book. Website links are for informational purposes and should not be seen as a personal endorsement; the same applies to the products detailed in this book. The reader should also be aware that although the web links included were correct at the time of writing, they may become out of date in the future.

Disclaimers

The content contained within this book is for information and entertainment purposes only, and in no way purports to represent professional medical opinion. It should NOT be used as a substitute for expert advice, and you must consult with your designated health professional before acting upon any information contained herein or before undertaking any practice whose methodology is referred to in this book. The author is NOT a registered health professional and the text merely represents personal opinion, not medical fact. The author cannot be held responsible for the consequences of any action derived from the reading of this book, as the content is not based on diagnosis and subsequent regimen. It is the reader's responsibility to seek proper, professional medical advice from a registered health practitioner in connection with any material contained within this book.

Legal Disclaimer (part 1)

Nothing in this book should be construed as an attempt to diagnose, treat or cure. The information in this book is intended to be a community resource. The author takes no responsibility for any informational material or brochures produced using information taken from this book. The author has endeavoured to ensure that all information is correct at the time of publication. This information, however, is subject to change without notice. The author makes no warranty with regard to the accuracy of any information and will not be liable for any errors or omissions. Any liability that arises as a result of this information is hereby excluded to the fullest extent allowed by law.

This information should not be used as a substitute for seeking independent professional advice.

Legal Disclaimer (part 2)

Disclaimer and Terms of Use:

a) i. In publishing this information, the author makes no representations concerning the efficacy, appropriateness or suitability of any products or treatments. Use this information at your own risk. The compiler is not a doctor and has no medical background or training.

ii. Statements and information regarding dietary supplements, books and any products mentioned have not been evaluated by any health authority and are not intended to diagnose, treat, cure or prevent any disease or health condition.

b) In view of the possibility of human error, neither the author nor any other party involved in providing this information, warrant that the information contained therein is in every respect accurate or complete and they are not responsible nor liable for any errors or omissions that may be found or for the results obtained from the use of such information. The entire risk as to use of this information is assumed by the user.

c) You are encouraged to consult other sources and confirm the information.

d) The information you access is provided "as is". No warranty, expressed or implied, is given as to the accuracy, completeness or timeliness of any information herein, or for obtaining legal advice. To the fullest extent permissible pursuant to applicable law, neither the author nor any other parties who have been involved in the creation, preparation, printing, or delivering of this information assume responsibility for the completeness, accuracy, timeliness, errors or omissions of said information and assume no liability for any direct, incidental, consequential, indirect, or punitive damages as well as any circumstance for any complication, injuries, side effects or other medical accidents to person or property arising from or in connection with the use or reliance upon any information contained herein.

e) The author is not responsible for the contents of any linked site or any link contained in a linked site, or any changes or update to such sites. The inclusion of any link does not imply endorsement by the author. The author makes no representations or claims as to the quality, content and accuracy of the information, services, products, messages which may be provided by such resources, and specifically disclaims any warranties, including but not limited to implied or express warranties of merchantability or fitness for any particular usage, application or purpose.

f) The information provided is general in nature and is intended for educational and informational purposes only. It is not intended to replace or substitute the evaluation, judgment, diagnosis, and medical or preventative care of a physician, paediatrician, therapist and/or health care provider.

g) Any medical, nutritional, dietetic, therapeutic or other decisions, dosages, treatments or drug regimes should be made in consultation with a health care practitioner. Do not discontinue treatment or medication without first consulting your physician, clinician or therapist.

h) By reading this information, you signify your assent to these terms and conditions of use. If you do not agree to these terms and conditions of use, do not read/use this information. If any provision of these terms and conditions of use shall be determined to be unlawful, void or for any reason unenforceable, then that provision shall be deemed severable from this agreement and shall not affect the validity and enforceability of any remaining provisions.

i) The information, services, products, messages and other materials, individually and collectively, are provided with the understanding that the author is not engaged in rendering medical advice or recommendations.

j) The information and the terms of use are subject to change without notice. The material provided as is without warranty of any kind and may include inaccuracies and/or typographical errors. The author makes no representations about the suitability of this information for any purpose. The author disclaims all warranties with regard to this information, including all implied warranties, and in no event shall the author be held liable, resulting from, or in any way related to, the use of this information.

k) The unauthorized alteration of the content of this information is expressly prohibited. The author, its agents and representatives shall not be responsible for any claims, actions or damages which may arise on account of the unauthorized alteration of this information.

9 798223 627784